Lecture Notes in Biomathematics

Managing Editor: S. Levin

24

Fred L. Bookstein

The Measurement of Biological Shape
and Shape Change

Springer-Verlag
Berlin Heidelberg New York 1978

Author
Fred L. Bookstein
Center for Human Growth
and Development
University of Michigan
Ann Arbor, Michigan 48109/USA

Library of Congress Cataloging in Publication Data

Bookstein, Fred L   1947-
   The measurement of biological shape and shape change.

   (Lecture notes in biomathematics ; 24)
   Bibliography: p.
   Includes index.
   1. Morphology--Mathematics. 2. Body size--
Measurement. I. Title. II. Series.
QH351.B66   574.4'01'84   78-15923

AMS Subject Classifications (1970): 41 A 05, 41 A 15, 41 A 63, 50-02, 62 P 10, 65 S 05, 73 P 05, 92 A 05

ISBN-13: 978-3-540-08912-4   e-ISBN-13: 978-3-642-93093-5
DOI: 10.1007/978-3-642-93093-5

2141/3140-543210

ACKNOWLEDGEMENTS

I have been the recipient of more diverse academic cooperation
than ever I expected. At Harvard in 1973, Raymond Neff (Biostatistics) underwrote preliminary sketches of this project, while Nathan
Keyfitz (Sociology) and Stephen J. Gould (Geology) offered early en-
couragement. At the University of Michigan, the Society of Fellows
granted me three full years of support as Junior Fellow, unencum-
bered by any responsibilities; William Ericson (Statistics), direc-
tor of the Statistics Research Lab, allotted me generous amounts of
space and computer time; Arnold Kluge (Zoology) remodelled draft after draft and provided audiences for my first trial runs at public
persuasion; Waldo Tobler (Geography) offered gentle competition and
shared computing lore; and Edward Rothman (Statistics), Gerald Smith
(Museums), George Nace (Biological Sciences), Philip Gingerich (Geology), Carl Simon (Mathematics), Robert Moyers (Human Growth), Leslie
Kish (Sociology), and William Durham (Human Ecology) all listened
hard and provided crucial suggestions.

Much of this essay was submitted to the Horace H. Rackham School
of Graduate Studies, the University of Michigan, as a dissertation in
statistics and zoology jointly. Some of the text of the second part
appeared in the journal Mathematical Biosciences, volume 34, and is
used by permission of Elsevier-North-Holland, Inc. The computer program which provided the geometrical examples here was produced in
the fall of 1977 with support from N.I.D.R. grant DE03610-11 to R. E.
Moyers and N.S.F. grant SOC77-21102 to the author. The other illustrations were drawn by Ms. Teryl Schessler.

                                        Fred L. Bookstein
                                        February, 1978

TABLE OF CONTENTS

TABLE OF CONTENTS (continued)

LIST OF FIGURES

LIST OF FIGURES (continued)

CHAPTER ONE.   INTRODUCTION:
ON THE ABSENCE OF GEOMETRY FROM MORPHOMETRICS

Let me refer to the craft of measuring biological objects and
phenomena as biometrics. In this century there have been three great
methodological inventions indigenous to biometrics, not borrowed from
outsiders. The analysis of variance, whose basic tenets Ronald Fisher
laid down to make easier the task of inference from biological experiments, has been transmuted into a powerful generalization, the
General Linear Model, and applied in virtually all fields of scholar-
ship. It is now considered a proper branch of statistics, and its
roots are lost sight of. Population genetics was formalized by
Fisher, Sewall Wright, J. B. S. Haldane, and others, beginning in
the 1920s, as a mathematical framework for discussion of populations
of genes changing over time. It has become entwined (in barbarized
version) in the current debates over I.Q. and in the technique of
path analysis lately popular in the social sciences. Quantitative
phyletics is relatively new and not so much the creation of famous
innovators. It has arisen to guide construction of the most crucial
formalism in modern evolutionary theory, the cladogram or chart of
evolutionary relationships, out of the chaos of contemporary and
surviving data.

These three great inventions, notwithstanding their various suc-
cesses in applications outside biology, share a common methodological
core. They address themselves to biological measurement problems,
generally disdaining models or techniques borrowed from earlier work
in other applied sciences. When they themselves have been borrowed
(as "anova" was), it is because the model originally invented to describe the biological phenomena (here, the statistical summation of
effects) has proven more or less applicable elsewhere; but this is
irrelevant to their success in biometrics. The alternative to this
invention of methodologies is their importation from other fields.
The major instance of this is modern multivariate statistical analysis, whose models have a wholly different flavor; they correspond to
exigencies of information processing generally, not to the ways of
biology, and are to be thought of more as tools for compressing measurements than as measurement techniques per se.

Today, biometrics is in need of a fourth indigenous development,
for the measurement of shape. This need is new, a consequence of
the ramification of biomedical image-processing technology (a spin-
off from military and space video research) over the last ten years.

For decades, of course, biologists have had access to photographs and diagrams as well as objects themselves in hand; but it was very difficult to extract interesting numbers. The best one could do then was measurement (by the painstaking application of ruler and calipers) of certain lengths, certain angles, and perhaps area or volume.

This is an unnatural hobbling of the range of mathematical models, restricted as it is to a geometry of lines and circles, a grammar unchanged since Euclid's time. Geometry has undergone several reconstructions since then, generalizing and broadening its sway. Many new constructs have entered--notably, for our purposes, differential elements, curvature, and convexity--and also several elegant new algebraic models for geometric entities, in particular the method of homogeneous coordinates and the tensor calculus. In comparison with current work, all these are considered old stuff, developed between 1700 and 1900; nevertheless, they are unfamiliar, in the main, to the modern applied scientist. The constructs of post-Hellenic geometry are basically more resourceful than the Euclidean. They involve assignment of numbers to all sorts of systems and aspects, not just lengths and angles. The basic discovery was Descartes'--that points of the Euclidean plane can be wholly characterized by coordinate pairs, and curves by equations. In the centuries since, mathematics has embraced this idea and extended it, to talk of the varieties of coordinate systems themselves, of change of coordinates and of the ways of combining or integrating numbers into invariants that do not change when the coordinates change.

This mathematical context underlies the possibility of a new morphometrics today. The necessary computations may be subtle and difficult to manage by "hand"--by ruler and other hand tools and desk-calculator formulae--but they are all more or less straightforward for the computer representation of a biologically relevant image. And the computation of _any_ structure or number which has a geometrical meaning and which serves to discriminate among shapes in important ways is as fairly called "measurement" as is the computation of distance using a ruler.

Because geometry itself has fallen from the twentieth-century curriculum, the ordinary scientist's geometric imagery is limited to the Cartesian tools he was taught in high school. For the more sophisticated measurement of shape there are simply no precursor applied fields to borrow from. When geometry has been applied, as in engineering, it is to the depiction of structures we know, single structures whose measures (usually physical quantities) are pre-assigned. Geometry here reduces to technical drawing. The reverse procedure, the passage from spatial structure to quantity, is crucially bound up with the scientific purposes of measurement: given a collection of similar objects (in the present study, similar shapes), to describe their variation and to relate it to other variations. This will prove much more than a matter of finite distances and angles. The problem here is indigenously biometric, for we know that shape, however measured, is biologically real and relevant, just as are yields and gene frequencies and evolutionary trees. As there is no other field in which shape is such a manifest variate, there is no other field from which to borrow a grammar of shape. The interested biometrician must revert to the spring of his mathematical models, to the various branches of classical geometry, and learn there all the ways of describing shape that might lead to useful, discriminating measurements.

My ambition in this essay is the rectification of the growing imbalance between biomedical shape data, fully incorporated in the computer age, and the archaic tools of morphometrics bequeathed us by the Greeks. I intend a thoroughgoing redefinition and reconstruction of morphometrics--the measurement of shapes, their variation and change--as a branch of applied modern geometry. I will try to set up geometric formalizations of shape which can serve as frameworks for quantification, for measurement of actual shape phenomena. I shall expound the geometry and outline the computations for the two of the procedures I have ascertained which measure shapes directly: the method of tangent-angle analysis and the method of skeletons. Then I will turn to the mathematically more intricate problem of measuring shape change, which has lain dormant since the flourishing of D'Arcy Thompson, and display in detail a method, a computer implementation, and worked examples. I fervently hope that my reformulation of the field will spur biomedical research to consider shape seriously as a variate capable of systematic measurement and suggest new approaches to all manner of outstanding problems and data.

FIRST PART.   THE MEASUREMENT OF BIOLOGICAL SHAPE

CHAPTER TWO.   SHAPES AND MEASURES OF SHAPE

If we are to construe morphometrics as the measurement of shapes, we need to know what shapes are and how we get them into position to be worked on by our tools.   Our appreciation of shapes is conditioned by properties of the Euclidean space in which we live with them and by their mode of biological production.

A.   Properties of the Euclidean Plane and Euclidean Space

The Euclidean plane, E^2, is a two-dimensional vector space over the real numbers in which distance-squared goes as the sum of the squares of the coordinate differences--the usual Pythagorean formula. Euclidean space, E^3, is a three-dimensional vector space over the real numbers with the same distance formula (now summing over three differences rather than two).   The implications embedded in these definitions have occupied mathematicians for more than two thousand years (Lanczos, 1970).   Alternate versions, all equivalent, define Euclidean space as a species of some more general space (affine, projective, "non-Euclidean," or Riemannian) upon which certain further identities are postulated having to do with perpendicularity (Pickert, Stender, and Hellwich, 1974), circularity (Klein, 1927: 133), parallelism (Lanczos, 1970:63), or curvature (Stoker, 1969:304).

From this variety of characterizations flow several of the properties I use in the sequel.   Simple (non-selfintersecting) closed curves divide the rest of E^2 into an inside and an outside--this is the Jordan curve theorem--and similarly simple closed surfaces divide E^3 (Spanier, 1966:198).   If we append to E^n the so-called points at infinity, there exist polarities, one-to-one involutory associations between points and lines; these give rise to the conic sections with all their familiar properties (Pedoe, 1970: ch. ix).   Smooth curves (twice-differentiable maps of a patch of the real line into E^2) have tangent lines to which they lie close; smooth surfaces (twice-differentiable maps of a patch of E^2 into E^3) have tangent planes to which they lie close.   Such curves have arc-length, and such surfaces have surface area.   If the rate of turning with respect to arc-length be given for the tangent vector to a plane curve, the curve can be reconstructed uniquely up to a congruence, a rigid motion of the plane (Stoker, 1969:27).   In E^3, surfaces are characterized uniquely up to

a congruence by the distances in various directions along the surface and the way the surface falls away from its tangent planes
(ibid.:138). Unless patches of surface are "flat," i.e., gotten by
rolling up a piece of paper into cylinder or cone, they cannot be
treated as equivalent to patches of the Euclidean plane, for distance
will not obey the Pythagorean formula whatever the coordinate system.
Gaussian curvature is a measure of this failure of the familiar formula, of the failure of circumferences of circles on the surface to
be proportional to their radii; it governs the rounding-up or more
complex contortions of surfaces in space (Hilbert and Cohn-Vossen,
1952: sec. 29). A closed surface on which all closed curves can be
shrunk to a point is called a surface of genus zero; it may be thought
of as a distortion of the sphere--unlike a doughnut, it does not have
any holes. For such a surface, the integral of the Gaussian curvature over the whole area is exactly equal to the constant 4π (Stoker,
1969:237). Such closed surfaces can have no global coordinate system without a singularity somewhere (ibid.:247). E^2 has one very
special property: it can be made into a commutative field by the
device of complex numbers, linear combinations of the x-axis and the
y-axis in the ratio of $\sqrt{-1}$. "Analytic" functions, functions over this
field which have well-defined derivatives, are the subject of a theory
which is perhaps the most exquisite achievement of nineteenth-cen-
tury mathematics. Certain of its discoveries shall be invoked in
the course of the present work. There is no equivalent for E^3.

We are especially interested in similarities, transformations
of E^n which preserve all ratios of distances. In E^2, every similarity is either a translation or a spiral similarity (rotation about
some origin with multiplication by a constant of all distances from
that origin), or one of these followed by a reflection. In E^3,
every similarity is either a screw displacement (rotation about some
axis with motion along the axis), a glide reflection (reflection in
a plane with motion along a line in the plane), or a spiral similarity (rotation about some axis, motion along the axis, and enlargement of scale) (Coxeter, 1961: secs. 5.4, 7.6). This last canonical
form is made the basis of David Raup's theory of mollusk shell construction (Raup, 1961). If a spiral dilatation brings an object
into exact registration with a part of itself, then its "shape" is
any connected plane curve which by iteration of the spiral dilatation
sweeps out the whole.

B. Outlines and Homologous Landmarks

A definition of shape begins with the notion of outline. An outline is a closed curve in E^2, or surface in E^3, which is twice-differentiable "nearly everywhere." Formally, this corresponds to the mathematician's notion of a closed chain of manifolds-with-boundary (Spivak, 1970: ch. 8). In a chain, every boundary submanifold is used twice. For closed polygons in E^2, every corner has two arcs abutting at it; for polyhedra in E^3, every edge is used by exactly two faces. One may think of an outline as roughly the image of some polygon or polyhedron under a smooth distortion of the space, E^2 or E^3, in which it is embedded. This definition excludes certain point sets which are better described by indices of texture and density-- sets like and and --and also some others which do not have a well-defined inside--sets like and .

To the corners of the polygon correspond points on the continuous outline where there may be no single tangent vector, but rather two limits of tangent position, one on either side, which need not be the same. Elsewhere than at these corners, the tangent vector of the outline must turn differentiably as a function of arc-length. In E^3, the situation is similar. An edge point is a limit of two paths of regular points (points where the normal rotates differentiably with respect to surface position) along which the normal tends to two different limits. Edge points lie upon space curves, informally called "edges" or "seams." Seams intersect at vertices, or apices, where they themselves are not smooth and at which intersect more than two paths with normal planes tending to different limits.

In biology there are no true corners, edges, or vertices, as everything is rounded in the small; but it will often be convenient to model observed outlines by constructions with seams or apices. In fact, no biological objects actually exist in an E^2--all are to some extent in the third dimension--but our data are often limited to projections, outlines, or other pictorial reductions, and our measures are much simpler if we ignore the third dimension instead of trying to reconstruct it or adjust for it somehow.

Corners and vertices of the outline correspond to special points associated naturally or perceptually with a biological form. In the field of craniometrics these are called landmarks, a term which I would like to allocate to general use here. They are presumed to be "homologous," that is, fully comparable in all their histological and topological characteristics, from specimen to specimen.

In E^2, landmarks can be conveniently divided into two categories, which I shall call "anatomical" and "extremal." Anatomical
landmarks are defined by some biological differentia. In a sagittal
x-ray of the human skull, there are such anatomical landmarks as the
sutures (or, rather, their passage over the crest of the skull), the
socket of the mandible, the cusps of the incisors, the brow ridge,
and many other structures which project onto very small sectors of
the boundary. Extremal landmarks are inferred from the geometry of
the situation rather than the biology. This category includes most
of the conventional cephalometric landmarks, but not all are invariant under similarity transformations, a demand we shall insist upon.
For instance, the pogonion is conventionally the frontmost point of
the chin--but the concept of "front" is not intrinsic to the shape
of the skull, and so the pogonion is geometrically not well-defined.
Other geometric landmarks, corresponding to points of high curvature,
tips of structures and near-corners, can be defined without reference
to particular Cartesian coordinate systems and are thus admissible.
Extremal characterizations involving distance are acceptable too:
points furthest from landmarks already well-defined, ends of the
greatest diameter of the form, and the like. Landmarks are most informative when they correspond to corners or to points of very high
curvature. When they lie instead on shallow arcs they can be replaced by "typical" points in their vicinity.

In E^3 the possibilities are, as usual, more diverse. I refer
again to the human head for my illustrations. There are, once again,
anatomical points--the incisors, the centers of the orbits, auditory
meati, foramina through bone. There are two sorts of extremal points
of curvature: positive extrema, corresponding to corners and to tips
of substructures, and also negative extrema, in crotches or saddle
points. Many other traditional extremal points of the cephalometric
literature are coordinate-dependent--the top of the head, for instance, which is called the acrion--and must be declared inadmissible
here. In E^3 there is also the possibility of extended boundary landmarks, anatomical lines, such as sutures, lips, hairline, and the
like; these lines may in turn contain landmark points anatomical or
extremal, just as the lips have corners. There is room in the classification for extremal lines also, where a shape boundary is very
nearly an edge, which may or may not connect vertices, as in the
sagittal crest of the male gorilla. In E^3, the most informative
landmarks are at vertices or points of very high Gaussian curvature.

Next most useful are points on crests or grooves, edges of one very
high principal curvature; least helpful are regular points, of both
curvatures low, whose location does not add much information to the
local form at all. These latter two categories of points have one
or two degrees of freedom for "typicality"--they may be replaced,
without significant loss of precision in locating the other ordinary
points of the shape, by points on a one- or two-dimensional locus in
their vicinity.

Landmarks do not define the form of any edge or surface. They
merely provide fixed points of reference upon it. To represent a
curving manifold we need an adequate sample of regular points (points
of continuously turning tangent or normal) upon it, whereupon we
reconstruct the continuum in between by selecting from some finite-
dimensional family of available forms. The regular points need not
be systematically measured from instance to instance of the shape;
that is, they need not be landmarks at all. The parameterizations
of the smooth forms can often be expressed in terms of a lattice of
locations, but this grid is not immanent in the resulting form or

measures. The dots in the diagram of a curve or a surface,

or   , merely reflect a particular local coordinate system,

and we are not concerned with that.

C. Definitions of Shape, Shape Change, Shape Measurement

Any outline in E^2 is mapped by the similarity transformations
of E^2 onto a four-parameter family of other outlines. We can trans-
late it to any origin (which is two degrees of freedom), rotate it
(one d.f.), or alter its scale (one d.f.). In E^3, the similarity
transformations map any outline onto a six-parameter family of other
outlines. (Translations now have three degrees of freedom, rotations
two, scale one.) We may define two outlines as equivalent if one
can be mapped onto the other by a similarity transformation in the
appropriate space, E^2 or E^3. As the similarity transformations form
a group, the equiform group, we may construct equivalence classes
of outlines, which are collections of all the outlines that can be
mapped one onto another by similarity transformations.

I define a shape in E^n as an equivalence class, under the equiform group, of outlines in E^n. Informally, a shape is an outline-
with-landmarks from which all information about position, scale, and
orientation has been drained. A shape change is a map of one shape

onto another which sends arcs (or surface patches) smoothly onto arcs and corners (or edges) onto corners; then it sends landmarks onto landmarks. A shape measurement is a function on some domain into the real line which is the same for all elements of an equivalence class. Informally, it is a function embodying the outline from which all information about position, scale, and orientation has been removed. The domain of the function is specified in various ways depending on the nature of the shape variation to be studied. Sometimes it will be a finite point set, sometimes the real interval (0,1), sometimes a more complicated structure, like the skeletal graphs of ch. iv.

This last definition is quite divergent from the current custom. In the biometric literature, shape is considered a matter of "shape variables," which are ratios of sums of distance measurements among landmarks (Mosimann, 1970, 1975; Sprent, 1972). These are among the ratios which are invariant under similarity transformations, but measures of shape are not to be limited to them. To the statistical analysis of the real-valued functions embodying shape, considerations of finite sampling and shape variable construction must apply; but the actual data are fully rounded form, not those lists of ratios. Where Mosimann speaks of shape with respect to a finite set of measures, I would insist that the problem is rather the extraction of a finite set of measures for shapes already given, out in the world. In particular, the conventional construction uses no information except the Cartesian coordinates of the corners, and that only indirectly, by way of interpoint distances. I constructed my definition to highlight the differential properties of shapes, the fact that they have infinite degrees of freedom. The direct construction of shape variables from distances does not allow such useful concepts as tangent angle and curvature (plane, principal, or Gaussian), and thereby no natural representation for bulges, bumps, and other immediate features of form. The limits of the conventional approach are thus fundamental; they are a consequence of proceeding to a finite set of shape variables without a construct for the shape itself. The basic analytic question, of the proper representation of curved form in a particular application, is thereby begged wholly and goes unexamined; the actual structure of E^n, the space in which the objects of study reside, is made moot.

Measures of shape appearing in other scientific literatures usually treat of polygons in the plane, point sets whose only "landmarks"

are their vertices. There is no room in the schemes for special
points which are geometrically ordinary. Reviews of these literatures include Bachi (1962), Zusne (1970: ch. 5), Clark and Gaile
(1973), and Duda and Hart (1973: ch. 9). Such quantities as emerge
are based on polygonal approximations to outlines, their vertices
combined into various formulas of dimension zero in length: angles,
normalized perimeter, area, and higher moments. These formulas capture such aspects of shapes as "compactness," "elongation," "circularity," "symmetry," or "convexity." In spite of a few attempts at
rigor (e.g., Bunge, 1962; Rosenfeld and Kak, 1976: ch. 9), the quantities output are ad hoc, not based in any theory of underlying quantitative information of which the measures used take a sample. This
can lead to absurdities. Brown and Owen (1967) found that twelve
linear factors explained 87 percent of the variance in measures of
shape for a random sample of quadrilaterals--but the object space has
only four geometric degrees of freedom.

Any attempt to describe systematic variation of shape, which is
our concern here, comes up against the need to associate specific
point pairs in the comparison of outlines, as in my formal definition
a few pages above. Occasionally the shapes of a particular subject
matter permit such a construction post hoc. Hilditch and Rutovitz
(1969) and Rutovitz (1970) construct the centromere of a chromosome
by projecting its density upon the principal axis of its planar
scatter and identifying the antimode. The location of this landmark
provides them with a crucial ratio for discriminating among the 23
possible shapes in the human karyotype. Widrow (1973), likewise attempting to identify human chromosomes, fits a "rubber mask" to an
observed form, a generalized letter H with nineteen parameterized
degrees of freedom for bending, twisting, and differential stretching.
An alternative procedure allows the outline itself to generate "landmarks" according to extremal properties which, however, fail to as-
sure homologies from case to case. For instance, Attneave and Arnoult (1956) suggested breaking up outlines at points of curvature
zero and infinity and at points of curvature derivative zero. The
resulting arcs are to be replaced by straight lines and arcs of circles. Systematic variation in the shapes of a subject field would
be replaced by systematic variation in these piecewise homogeneous
replacement curves. This approach has been generalized within the
field of picture processing to a variety of algorithms for replacing
empirically determined curves with straight lines only (Pavlidis and

Horowitz, 1974; Freeman and Davis, 1977), circles only (Shapiro and Lipkin, 1977), or straight lines and arcs of conics (Cooper and Yalabik, 1976). All such methods are flawed by the indeterminacy of that list of "homogeneous" pieces. The landmarks do not correspond one for one, neither do the lists. For statistical analysis, these lists of variable length must be replaced somehow by functions of constant domain.

D. <u>Shapes as Data</u>

A word is necessary regarding computations involving shape. The mathematical demands of the model must be modified somewhat in the encounter with actual data. We are, after all, confined to digital computers, where we cannot represent a continuously curving boundary in E^2 or E^3. There are two customary ways of adapting shapes to this milieu, that is, "inputting" them. Hursh (1976) surveys the history and variety of the machines which help to do this. We might measure two or three Cartesian coordinates for some points on the boundary: landmarks and few or many regular points in between. This can be done to considerable accuracy by machines which a human operator merely "points at" a series of useful locations, whereupon their picture coordinates are retrieved electronically (Walker and Kowalski, 1971); the third coordinate may be retrieved by use of a pair of pictures, in the technique known as stereophotogrammetry (Herron, 1972). The form of the curve or surface in between the points sampled is presumed some smooth construction going through all the data given, except, of course, where the landmarks represent explicit breakdowns of the smoothness criterion. (This is a mathematically ambiguous prescription whose particularization depends on the selection of some aspect of nonsmoothness to be minimized, usually higher derivatives or curvature. Cf. Schumaker, 1976; Barnhill, 1977.)

As an alternative to the exact input of a small point set, one might receive a whole lattice of cells, ordinarily square cells in E^2 (or cubes in E^3), coded for being "inside" or "on" the shape of interest. Such a presentation is prepared by a digitizer, which needs a rule for deciding, without "knowing" the shape, which points are to be considered inside. There are several ways of doing this. In chain-coding, a fairly clear curved boundary (perhaps traced by a human operator) is converted to an ordered series of lattice points, those it passes closest to. From any point on this chain one must proceed in one of exactly eight directions-- --and so the coding is quite compact (Freeman, 1974). In the threshold method, the digitizer sets to "inside" those cells of integrated optical density exceeding a certain threshold, or generates a "fuzzy boundary" by setting to "edge" any cell which looks like an edge, where the spatial gradient of density exceeds a certain threshold (Prewitt, 1972). This thick point set is then shrunk to a chain by clever software. The threshold method has been adapted to complex images of widely varying average level by searching for bimodal structure in histograms of optical density in middle-sized windows all over the image. A frequency trough between two peaks is deemed to indicate an edge and used to calibrate a local thresholding. There results a global separation of regions corresponding to no single dichotomy of the density function with which we began. Even landmarks can be represented in the course of this process, as the corners of lungs are traced out painstakingly over a chest x-ray; this provides a sort of feedback to the edge-tracing and smoothing processes both (Chow and Kaneko, 1972; Wechsler and Sklansky, 1977).

Except for these helpful suggestions about acquiring our data, the congeries of techniques which is "digital picture processing" is of surprisingly little use to the biometric investigations of this essay. The problem is not with the biological subject-matter but rather with our strictly metric intent. Conventionally applied picture analysis, as reviewed, for instance, in Rosenfeld (1976), is concerned in the main with classification, with "pattern recognition," and not with the particulars of quantification among objects whose general pattern is unvarying. The reader who wishes to browse in biometric aspects of that more general literature, in which, for instance, the topological structure of outlines is not assigned a priori, would best begin with the exhaustive bibliographies of Rosenfeld published annually since 1972 in the journal Computer Graphics and Image Processing.

CHAPTER THREE. CRITIQUE OF AN APPLIED FIELD:
CONVENTIONAL CEPHALOMETRICS

In this chapter I shall demonstrate the extent to which one particular field of applied shape measurement, conventional cephalometrics, is founded upon assumptions and procedures with unfortunate
analytic properties. A critique of certain modern developments will
be found in chapter iv, and of prediction techniques in particular
in chapter viii. There is a collective criticism of specifically
radiological cephalometrics in Salzmann (1961).

A. Landmarks, Curvature, and Growth
    In conventional cephalometrics, as reviewed by Krogman and Sas-
souni (1957), Howells (1973), or Merow (1975), the face and cranium
are measured by locating on the projected images a large number of
landmarks, special points with names and operational definitions.
We measure their locations in terms of distances between pairs of
landmarks, distances between landmarks and lines through landmarks,
and angles between pairs of lines through landmarks. The remarkable
antiquity of this strategy is reviewed by Gysel (1972).
    The available landmarks may be classified as "anatomical" or
"extremal." Anatomical landmarks, identified by some feature of the
local morphology, include cusp tips of teeth, nasion, sella turcica.
Other landmarks, the extremal, are defined implicitly by the maximum
or minimum of some geometrical property. Unfortunately, the important extremal landmarks are usually defined in terms of a sample-
specific horizontal and vertical. The menton is the lowest point of
the mandibular symphysis; the pogonion is the most anterior point on
the chin, the gnathion the lowest most anterior point on the chin,
the condylion the most superior posterior point on the condyle. As
the mandible rotates, the positions of all extremal points are altered. Then such points cannot be located until an orientation, let
us say the horizontal, is fixed. Now, all the standard orientations
are themselves operationally defined in terms of landmarks too. The
Frankfurt plane, for instance, passes through top of porion and bottom of orbitale; but "top" and "bottom" are themselves defined in
terms of orientation to earth. Malplacement of the horizontal orientation may be due to a mistake in positioning the subject or to inappropriateness of the landmarks for that subject, perhaps because of
disproportionate growth or asymmetry. This makes the whole scheme
susceptible to a pernicious form of measurement error, in which error

in the orientation is propagated to affect the positions of all the orientation-dependent landmarks instead.

Even were there in practice no errors of orientation, I would argue that for analysis of form and the change of form any scheme involving only landmarks is inadequate in principle. Analysis of the various distances and angles in this tradition proceeds by way of their sample means and covariances. There results a structure of inference for the ordination of specimens and the comparison with "norms." I argue that no such linear analysis can apprehend true curving form. Consider, for instance, the three landmarks shown in Fig. III-1, which manifest an angle of 160°, $\angle ABC$. The true outline through these three points can vary with infinite subtlety around the landmarks, and the meaning of that angle changes with it. The figure shows three of the possibilities. These are very different forms--arc, bulge, wave, we might call them--but they all have the same angle, the same landmark description. In the study of growth or remodelling the problem is equally difficult. If we are told that the angle $\angle ABC$ in Fig. III-2(a) increases, we can decide among the possibilities (b), (c), (d) (and others) by auxiliary information on distances. The curves of these figures correspond to quite different states after the next increment of time, so it is crucial that we discriminate them. But Fig. III-1 can be read, too, as a history of form change in which the landmarks have not moved at all; and these, too, correspond to future states of considerable variety. How, then, can we hope to gain information about growth and remodelling from the landmarks alone?

Any collection of individual landmarks, i.e., uniquely named points, obscures the continuous variation of form in between. The form is not, as in Fig. III-3(a), delineated somehow by all possible segments between pairs of landmarks, and the associated collection of all possible distance-ratios and angles; it is delineated rather as in (b), by the curving of the outline around the whole. The head contains no polygons. Landmarks, progress-markers on the circuit of the form, do not define it, but merely lie upon it. In particular, growth is everywhere between the landmarks, not localized at them in any sense. To understand the changes of growth and remodelling, we need to know how each landmark is moving away from the others. It would suffice even to know how each is carried away from those on either side, changing local distances and directions simultaneously. Such an analysis is necessarily formulated in terms of the curved form in between landmarks, which is altering the relationship of any

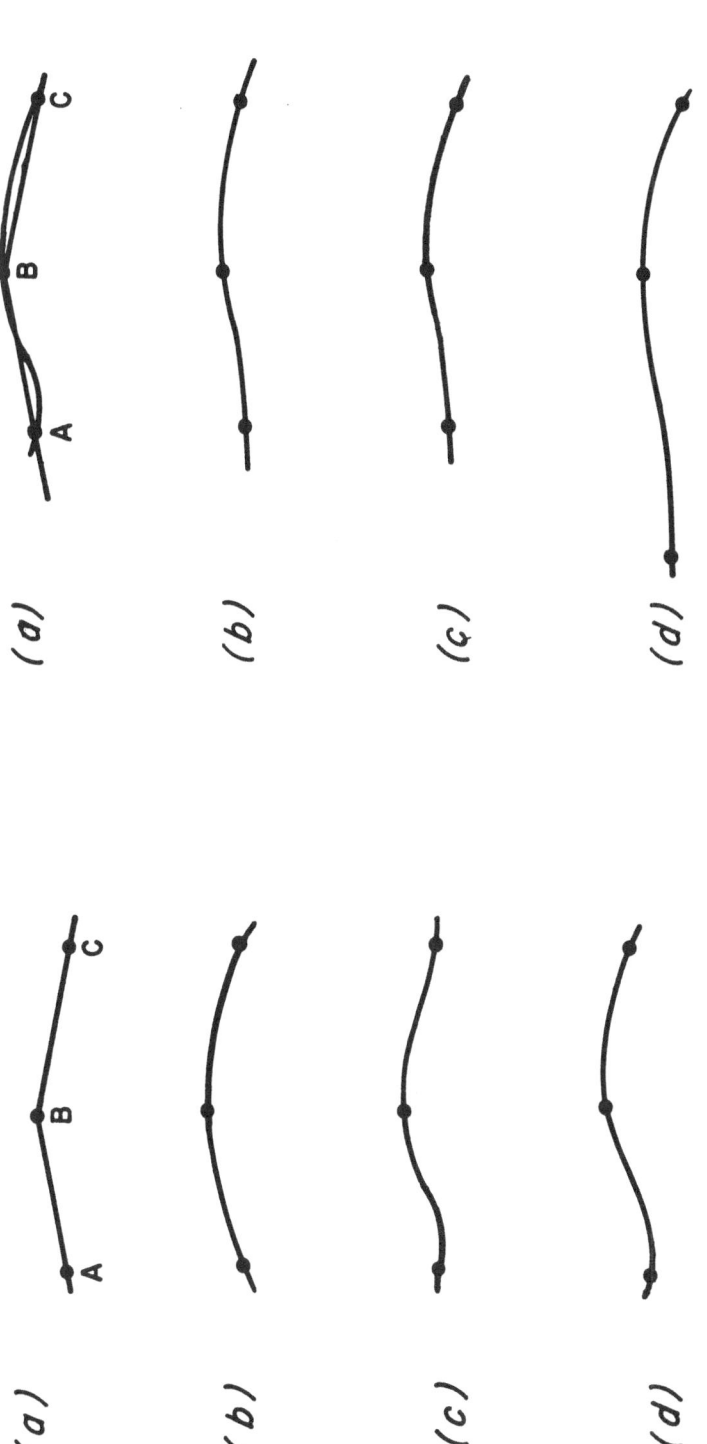

Fig. III-1.--(a) Polygonal geometry of a configuration of landmarks. (b, c, d) Three outline forms: "arc," "bulge," "wave."

Fig. III-2.--Three ways in which the landmarks of Fig. III-1(d) may have moved to increase angle ∠ABC. (a) Starting position. (b, c, d) Point B drops; point C rises; point A moves aside.

(a)

(b)

(c)

Fig. III-3.--(a) The totality of lines and angles available from constructions on landmarks only. (b) Additional information at each landmark, omitted from the representation in (a). (c) Geometric meaning of tangent angle and curvature at landmarks.

pair of landmarks in a geometrically intricate fashion.

If we now speak of actual quantitative data extracted from cephalograms, then the indices associated with Fig. III-3(a) are not very
useful for analyzing change, for they do not even distinguish the
situations of Fig. III-1. We could take a giant step toward adequacy.
of measurement by noting, as suggested in Fig. III-3(b), the tangent
angle and curvature in a little neighborhood around each landmark,
in addition to the landmark position itself. The tangent angle is
the azimuth of a straight line lying along the outline at the landmark,
as in (c); the curvature is the inverse of the radius of the circle
closest-fitting to the outline there, and indicates how rapidly the
tangent angle is varying with distance along the outline. It is precisely these two measures that are embodied in the little hachures
which make (b) so much more informative than (a).

No extant cephalometric scheme includes these two simple measures, for they arise mathematically as limits from the construction
of segments through pairs and triples of points arbitrarily close and
so override the basic notion that landmarks should be individual loci
at some finite spacing. (There are in the literature an assortment of
hybrid measures which estimate curvatures along arcs by reference to
preset chords. These measures depend crucially on the chords selected, and cannot be generalized to continuous metrization. For
instance, in all the many indices which Young, 1956, assembled for
net cranial bulging between nasion and bregma, the bulge at either
landmark is necessarily exactly zero.) Nevertheless, the data required for computation of figures like (b) already exist. Tangent
angle and curvature can be estimated quite easily from digitized
cephalogram tracings of the sort described by Riolo et al. (1974)
and Walker and Kowalski (1971, 1972). Hitherto these data banks have
never been processed in this way, but only for extraction of the old,
inadequate indices. There is no reason this limitation should continue.

Landmarks do not define the form, but only serve as pointers to
hold our conceptual place upon it. We need not be satisfied to measure tangent angle and curvature only at landmarks, then; we would
ideally want the history of these quantities all the way around. At
this juncture mathematics comes to our rescue with the handy fact
that the two parameters we find so useful are redundant. Curvature
is merely the derivative of tangent angle with respect to arc-length
along the curve (cf. chapter ii). The information we want is con-
tained in a single function, tangent angle, defined all around the

outline of the cephalogram. The sections of Fig. III-1 are then represented as in Fig. III-4, where they can be told apart quite trivially. Landmark data merely sample this function at points which are located reliably but which do not necessarily provide any especially useful information. To the extent that cephalometrics is based wholly upon landmark location and quite ignores the curving of the outline there and in between, it cannot possibly measure individual form in a thorough fashion.

When data do not capture the curving of form, the problem of classifying facial types over a succession of ages is especially difficult. The form of any face or part thereof changes because of differential growth rates in different directions throughout the craniofacies. The types an orthodontist assigns and uses, such as Angle

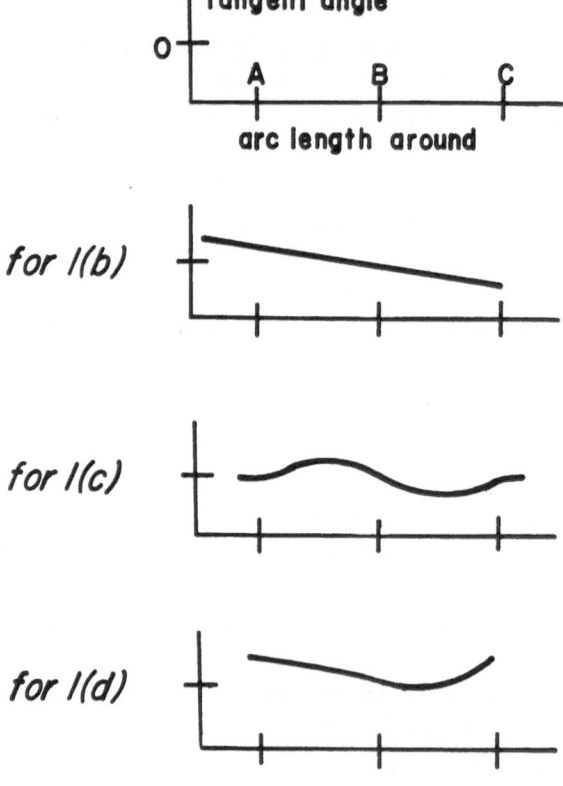

Fig. III-4.--Sketches of tangent angle as a function of arc-length for the three curves (b, c, d) of Fig. III-1. They can be discriminated very easily.

classes, should properly be constructed as invariant with respect to
the general run of growth. Now growth itself is not compatible with
the current cephalometric scheme, for the moving apart of landmarks
expresses diverse form change everywhere in between and is not at all
summarized by any set of distances and directions. The landmarks are
but carried along on a field of continuous spatial deformation. In
the current state of diagnosis, based on polygons, we have no way of
searching for structural relationships invariant under the normal
range of differential growth rates, for we have not measured these
rates, even indirectly. Heads and faces actually change by bulging
and curving, not just by displacement at the corners of polygons.
Any geometric construction on landmarks alone, for instance the intersection of two lines and its concomitant ratios, will itself be distorted over successive ages of even "normal" growth owing to idiosyncrasies of regional growth rate; and if the form begins abnormally
it will likely be distorted abnormally as well. In a situation such
as that of Fig. III-5, the distortion cannot be appreciated without
a material history of that central point of intersection--but it is
not likely a landmark at all, and goes unmeasured. In these circumstances the stability of quantitative classifications upon successive cephalograms would be a matter of the sheerest luck. We cannot
correct for growth rates, for we have totally failed to measure them.

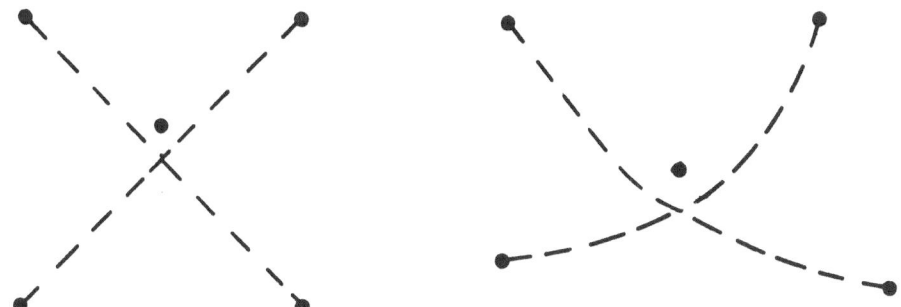

        Fig. III-5.--A typical distortion of growth, invalidating the
ratio in which one line cuts another as a useful indicator of change.

B. Registration
        Thus far I have been criticizing the cephalometric approach to
measurement of cephalograms singly, presuming that analysis of change
is the analysis of series of numerical measurements. There is a pop-

ular alternative for the study of change which does not commit the
fallacies of the previous discussion; it commits others instead. I
refer to the technique of superimposed tracing. Successive tracings
of cephalograms are placed upon each other according to somé rule,
and the motion of corresponding landmarks is plotted; or points of
a single cephalogram are each moved a bit according to some population average motion, and the resulting future state drawn out in line.
Such a line following the progress of a single landmark will here be
called a track. Fig. III-6 displays a collective cephalometric history by this method. In principle such a diagram evades the objections we presented for the previous technique. The "motions" of
landmarks are now just a sample from a continuum all the way around
the outline. The little tracks show how growth of all landmarks proceeds away from some central point of the skull, usually the sella--
from this can we not somehow infer the divergences between neighboring landmarks which, according to theory, are the real mechanism
of change?

Alas, we cannot. For any of these techniques, statistical
treatment is extraordinarily problematical. All location measures
are beset by the measurement error of the standard location and orientation in addition to their own. This supplies a functional correlation among any pair of displacement measurement errors which is of
constantly changing magnitude and direction. Points near the center
of registration move less than points farther away, and all points
generally move away from the fixed point--both these trends are anatomically quite meaningless. The divergence between neighboring landmarks is much less than their common translation, for when a bone is
translated by growth in one region, all landmarks on its far side
are passively moved in addition to growing on their own. Then the
structure of variation of the outermost tracks is a statistically
intransigent composite of variations there and elsewhere; to untangle
it one must start over by other analytic procedures.

Behind these statistical difficulties is an even more basic flaw.
The conclusions we draw necessarily vary, quantitatively and qualitatively, with the registration rule. The superimposition of two
images generally proceeds by specifying one registration presumed
unmoving and one direction presumed not rotating. What is drawn out
is a record of two quantities assembled in polar coordinates: the
distance from the point of registration to the landmark of interest
(which distance is a summation of rates and directions of all the
growth in between) and the angle among the landmark and the two points

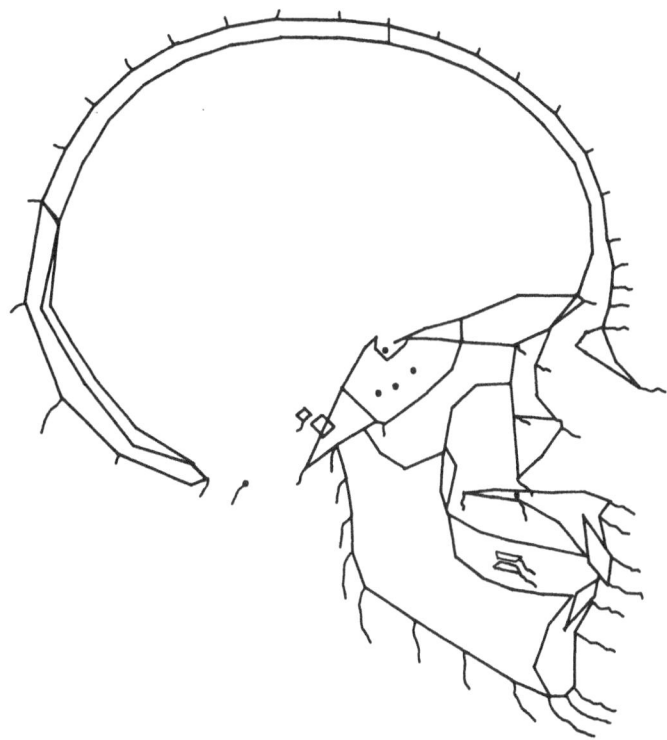

Fig. III-6.--A complete cephalometric history of a population of 60 females. The "whiskers" each track one landmark through a sequence of positions on successive cephalograms (only the first of which is drawn in full) which are stacked so that all the images of sella are superimposed and all the lines from base of occipital bone to center of palate are put parallel. After Walker and Kowalski (1972).

making up the line of orientation. Any revision of the orientation, then, and also any measurement error in the coordinate system, will change this track more or less drastically as the landmark is a greater or lesser distance from the fixed point. Depending on the orientation and registration we choose to hold fixed (Frankfurt horizontal, sella-nasion, orbital plane, or whatever), the progress of particular landmarks can appear to be curving up or curving down, speeding up or slowing down on their tracks.

Registration need not be upon measured point-landmarks at all. Björk and Skieller (1972) argue that for studies of craniofacial growth the jawbones ought to be registered separately upon certain

fixed structures, extended in space, which histologically show nearly
no growth at all. This registration is demonstrated by way of me-
tallic implants in the jaws, implants which do not move relative to
each other or the bony matrix. When successive images of the same
mandible are superimposed on the implants they carry, one sees (Björk,
1969) that the mandibular canals (pathways, visible in the x-rays,
through which pass the blood vessels and nerves supplying the teeth)
are themselves virtually invariant as remodelling proceeds around
them. Then normal mandibles, without implants, can be legitimately
superimposed by registration and orientation upon those canals, and
such phenomena as eruption of teeth may be measured over time in this
corpus-based coordinate system.

Such convenience does not obtain for landmarks not on the selfsame rigid structure. Those bones from which the mandible is pushed
away in growth, the maxilla and the temporal, must be measured from
their own fixed points, not the mandible's. The relation between any
two of these coordinate systems is quite complicated, as their points
of contact (incisors, molars, condyles) all grow simultaneously according to all coordinate systems. Such formalisms reduce the motions
of all parts of a growing system to the relative motions of rigid
bodies (the inter-implant segments), relative motions decomposable
in turn into translations and rotations (see chapter ii). There result diagrams of change as relative flexure of mechanical linkages
(cf. Marey, 1895, for spectacular examples). In this presentation
rounded form is suppressed, and with it fades all possibility of the
geometric specification and localization of growth. Translation and
rotation are characteristics of motion of rigid bodies only, not of
growing tissue.

It is also possible to superimpose statistically, on averaged
structures, for instance registering on the centroids of areas delineated by the data. Furthermore, ordinary least-squares regression methods may be adapted to the computation of a best-fitted relative orientation for landmark point data (for one technique, see
Sneath, 1967). In fact, registration, orientation, and a change of
scale (that is, the general Euclidean transformation) may all be
fitted simultaneously by the technique of Helmert regression (see
Tobler, 1977). For affine transformations, that is, general plane
shears, I cite some methods in chapter vi.

Now all these registrations contain <u>exactly the same geometric
information</u>--the motion of all points relative to a moving coordinate

system. We can pass from one to another by adjusting trajectories
to take into account the simple formulae for change of polar coordinates. The shape of the little curved tracks, then, has no biological
meaning whatsoever, for we can make any track go to a fixed point simply by registering directly upon it, without any loss of information.
There is then no best registration. That the information content
of alternative registrations is equivalent does not save us from statistical fallacies, inasmuch as all our statistical models are fundamentally linear. A track which is straight or in any way well-
behaved in one analysis will be curved and difficult to analyze in
another. This can be argued best by means of an artificial example.
Fig. III-7 shows three successive states of a quadrilateral of landmarks in a hypothetical developmental sequence. Selecting point A
for registration and the line AB for orientation, we obtain the
tracks in Fig. III-8(a); selecting point A and the line AC, we obtain

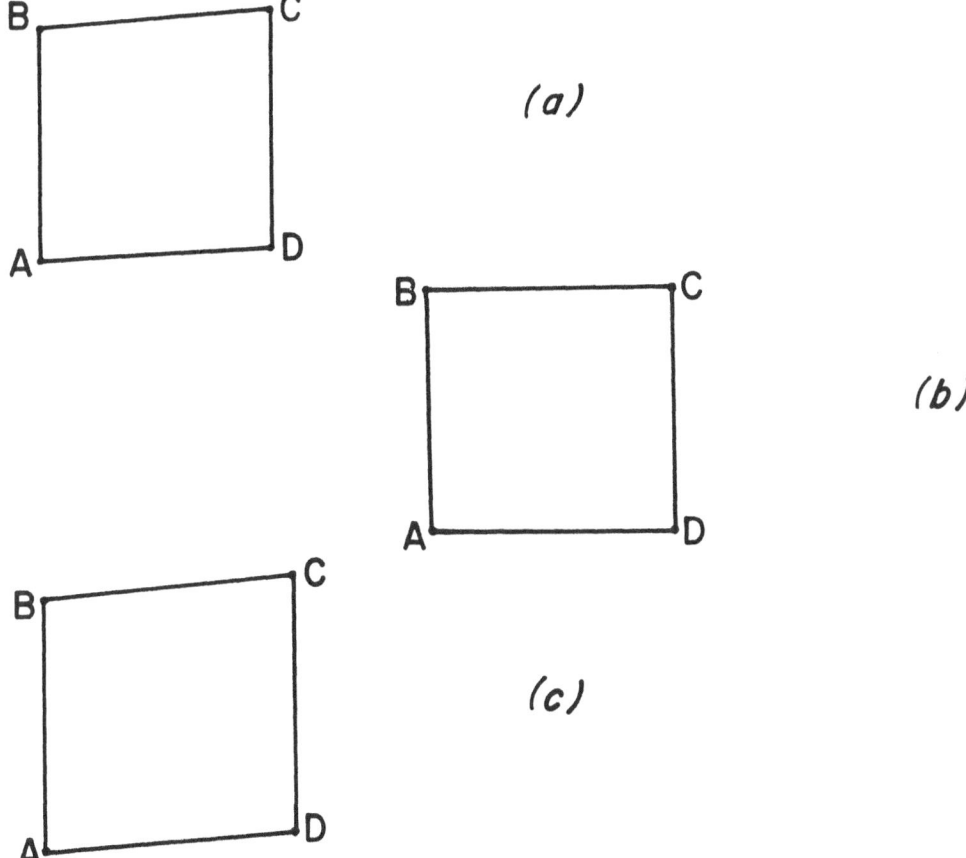

Fig. III-7.--A hypothetical ontogenetic sequence of three
cephalometric polygons.

(b); registering on A and orienting on line BD, we obtain (c). The raw data are unchanged, but the registrations are related to each other in non-linear ways. The simplest way of viewing these data is as in (d), oriented on lines AC and BD with their intersection for register. The shape change was in fact produced by an extension along line BD followed by an equal extension along line AC. This is not

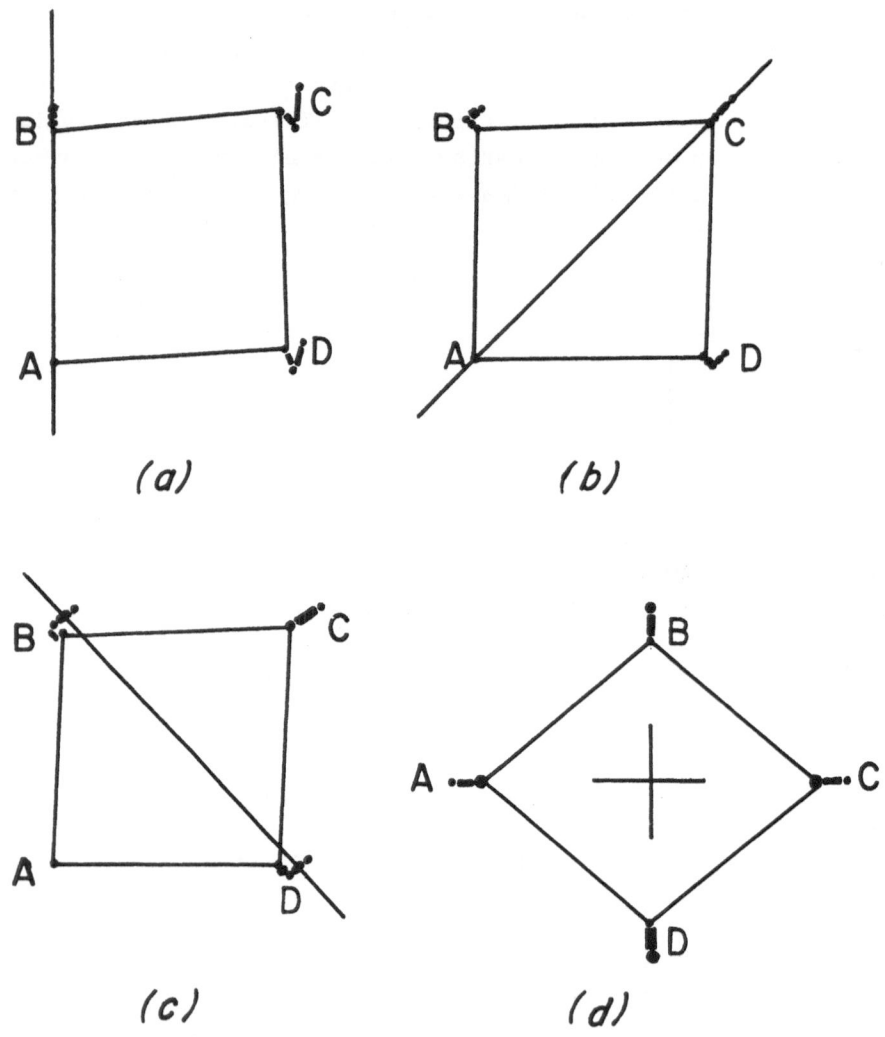

Fig. III-8.--Summaries of the history of Fig. III-7 by four different registration rules. The topmost outline of the history is drawn out in full, and upon it the trajectories of its corners, properly oriented and registered, are traced in broken lines. Only for the last tracing, which is not registered on a landmark, is the visual summary biologically suggestive.

so unreasonable a model for certain changes in the face--but how un-,
promising analyses (a), (b), (c) looked for the same data!

It is no accident that the registration for (d) is on a constructed point rather than a landmark. Any pattern in which the
points used for registration or orientation are involved will be
wholly misconstrued by registration or orientation upon them. Their
spatial variation has been arbitrarily restricted to a line or a
point, while all other points are free to vary over a region of non-
zero area. In particular, registration at sella with orientation on
sella-nasion is predicated upon the postulate that growth there is
functionally unrelated to change in the parts of the face that are
of orthodontic interest, that the geometric straight line from sella
to nasion does not geometrically bend over the course of growth.
Likewise, use of the Frankfurt horizontal requires linearity of
growth all along the line from porion to orbitale. In fact the crucial points of either registration do not move smoothly with respect
to the other registration, and certainly both systems are correlated
with developments elsewhere in the craniofacial complex.

How can any such assumption of "featureless" registration, registration not implicated in the conclusions we draw, be defended?
No known registration yields smooth linear tracks for the landmarks
distant from the fixed point. Growth rates vary irregularly in space
and time over the developing face and cranium. In the absence of any
metric theory describing how the landmarks are borne apart by the
growth in between, the legibility of any registration, the appearance
of consistency among tracks, is a matter of chance. As all registrations express the same geometric content, the preferred choice can
be only an accident of near-linearities over an extended region--
averages of nearly balanced non-linearities acting everywhere
throughout. This is the most evanescent of statistical advantages,
and in particular a registration which works for normal cephalograms
can be expected to fail systematically for cases with any notable imbalance. However we proceed, non-linearities always enter the data
analysis. The cephalometrician does not yet have any grasp of the
functional constraints relating sizes and rates of growth throughout
the face; therefore he cannot decide whether any registration is con-
cealing crucial covariation. There is no essential difference among
the alternate techniques, as all share this fundamental statistical
flaw.

Let me summarize this manifold critique. Conventional cephalo-metrics is inadequate for description of individual cephalograms ow-ing to its inability to apprehend curved form; it is limited to land-mark-based indices. The problem of recognizing stable patterns over time is especially difficult, as there exist local directional growth gradients all over the face which the technique cannot take into ac-count. For description of growth or shape change the usual techni-ques are all beholden to one or another systematic registration and orientation, which distort irrevocably the very record of change we would like to examine, and which have a systematic error interacting in an unknown fashion with particular abnormalities of form.

CHAPTER FOUR.   NEW STATISTICAL METHODS FOR SHAPE

In this chapter I propose statistical analyses of data which
preserve the full rich information store of the outline as late as
possible into the analysis. We can only deal ultimately, it is true,
with a finite-dimensional space of parameters; but when shape is
represented by a continuous function it is possible to sample from
that function instead of discarding information earlier by sampling
from the points of the outline. Distinctions between the methods I
shall describe, and between them and the conventional approaches, can
be summarized in the three schemes of Fig. IV-1. Sketch (a) indicates
the conventional method of extracting shape variables from landmark
locations alone, ignoring both the curving of the outline and the
geometric order of the image. Sketch (b) corresponds to the tangent
angle method, which follows the outline all the way around. Sketch
(c) is a useful modification, the medial axis method, in which we
pass up the "middle" of the form, in a sense to be defined objective-
ly, and investigate the boundary on both sides simultaneously.

A.   Analysis of the Tangent Angle Function
1.   History
Statistical analysis of shape functions going all the way around
a boundary has precedent in various applied sciences. I have encountered three embodiments of this intent in diverse fields. Each
bears embedded statistical flaws which disqualify the various techniques from more general domains of analysis.
I have previously mentioned the exposition of Attneave and Arnoult (1956), who noted that points of especially high (or especially low) curvature were informative to the human observer and useful
in the analysis of outline forms. In certain empirical applications
these extrema are reliably present, that is, they take on the formal
properties of landmarks. Ledley (1972), summarizing his work of the
1960s, represents the shape of a chromosome in terms of the maxima
and minima of curvature of a boundary represented by a chain. At any
point x, Ledley's algorithm approximates the radius of curvature of
the boundary by the radius of a circle tangent to the segments xx',
xx", where x', x" are boundary points within a few chain links of x
on opposite sides. The values of the radii so computable are a discrete set. This procedure is good for locating the extrema of curvature, which Ledley then uses to roughly reconstruct the arms of the
chromosome and thereby determine which of the 23 human forms it likely

Fig. IV-1.--Three ways of representing one outline. (a) Landmarks and lines through them. (b) Tangent angle and curvature at
the landmarks. (c) The skeleton, consisting of points which do
not have a unique nearest neighbor upon the outline.

is. There is no attempt at quantitative reproduction of the outline in between. The procedure is therefore not suited to more subtle quantitative investigations. See also Rosenfeld and Johnston (1973).

There have been several attempts to represent the entire curvature function around the outline in terms of its Fourier series (Zahn and Roskies, 1972; Bennett and Macdonald, 1975; Eccles, McQueen and Rosen, 1977; Persoon and Fu, 1977). The first few coefficients of the series are then used as "measures" of the outline for entry into descriptive and discriminatory computations, primarily for the recognition of printed or handwritten numerals and characters. There are several problems with the general application of this technique. There is room in the scheme for at most one landmark, where the series is presumed to "begin." Otherwise it can manifest no local features by way of the coefficients, for each term is an aggregate measure in both its amplitude and its phase. Furthermore, the true curvature function of all the outlines to which this method has been applied is a delta-function, that is, a function which has infinities of value representing sudden jumps of the integral. (This is because the data arise from chained representations.) With more and more terms of the series, convergence is to a jerky polygonal outline rather than the smooth underlying form we seek to measure. An alternative to Fourier analysis of curvature is the expression of the entire planar extent of the form as an infinite series in the Walsh functions, which are orthonormal series of the same formal properties as the trigonometric terms of a Fourier series. These functions have proved very useful in the numerical representation of whole images, but for statistical analysis of shapes they have exactly the same drawbacks as the method preceding. See Meltzer, Searle, and Brown (1967) and Searle (1969).

In a variant of this technique, P. E. Lestrel (1974), studying the hominid calvarium, executes a Fourier analysis not of the curvature function of the outline but of the radius length, measured from a point inside the form, halfway around the head. The generation of coefficients in this way was previously tried by Rutovitz (1970) without apparent issue. Lestrel proceeds further and submits these Fourier coefficients to a multivariate analysis, emerging with two principal components of the measurement vector. On a plot of one principal axis against the other he locates specimens and species. The technique allows two landmarks, one the starting point of the series, one the point chosen as center of coordinates. All the objections listed in the preceding paragraph apply to this method, and

also one further crucial difficulty. Each analysis depends fundamentally on the landmark used for the center of polar coordinates.
Error in the location of that point is expressed in an alteration of
every value of the radial function and therefore every coefficient
of the Fourier series, in a complex, non-linear way. Lestrel uses
the external auditory meatus in his examples. This is an area, not
a point; it is not symmetrical as a shape, so that it has no precise
center; it is hard to delineate on cephalograms; and it is asymmetric
from side to side of the cranium. The alternative of the centroid,
Rutovitz's solution, is no better. Which centroid shall we use--that
of the outline (a one-dimensional density) or of its interior (a two-
dimensional density)? Either location is a function of the very
shape we are trying to measure; this functional dependency completely
confounds the measurement process.

2. Sampling from the tangent angle function

I propose a more geometric method of greater generality. Assume we have in hand a particular closed curve with landmarks on its
outline. This shape, we know, can be exhaustively represented (i.e.,
without loss of any information) by way of its curvature or its tangent angle function all the way around. Figure IV-2 shows the representation of a familiar shape in this manner. I have argued in chapter iii that there is a great deal more information in this function
than in the conventional measures. The landmarks are milestones upon
this passage around, and thereby sample the function depicted. Their
actual Euclidean locations, up to the equiform transformations defining the equivalence class of all similar shapes, are implicit in the
function. We describe this sample by extracting both coordinates,
tangent angle and arc-length, of the points shown in black in the
figure. In a population of comparable shapes, our ultimate subject
of statistical study, both of these are variable quantities. If there
are N+1 landmarks, we can standardize the tangent angle function so
that arc-length runs from 0 to 2π (making arc-length and azimuth
commensurate) and so that the tangent angle at the first landmark,
where arc-length is exactly 0, is $0°$. Then the function is sampled
by N paired measurements (arc-length, tangent angle), forming a 2N-
vector. Each shape of a population of shapes provides one observation of this vector. I propose that such vectors be routinely extracted from shape data and submitted to the usual multivariate statistical analyses. Then regularly covarying features of the curving
outline will emerge in the course of inspection of loadings, just as

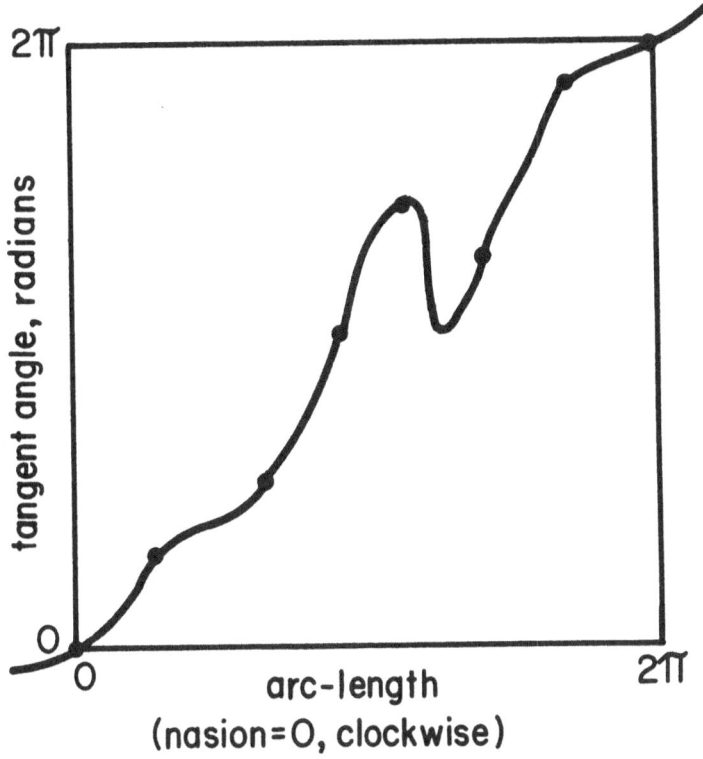

Fig. IV-2.--Sketch of the tangent angle function for IV-1(b), which is a schematic of a human head, in sagittal projection, ignoring the nasal region.

for conventional cephalometrics (cf. Howells, 1973) regular features of interlandmark distances and angles emerge from loadings. The representation of an outline by a sample from its tangent angle function avoids the problem of deciding which distances and which angles in the hugely redundant set of Fig. IV-1(a) to measure; it includes them all, implicitly.

Conventional multivariate analyses must be modified somewhat to take into account the geometric ordering of the data underlying these vectors. The results of any shape analysis ought to be diagrammable: there ought to exist a geometric realization of any coefficient vector (means or loadings) as a closed curve comparable with specimens of the original shapes. Most contemporary analyses do not have this

essential property. For instance, in a population of shapes reduced
to three landmarks and measured by four assorted indices, as in Fig.
IV-3, the population means for sides A, B, C describe a triangle
whose angle $\hat{a}$ is not, in general, equal to the mean of the observed
angles $\bar{a}$. In this situation the multivariate mean vector cannot be
diagrammed, for it does not satisfy the constraints of Euclidean geometry. In a multivariate analysis, a vector of loadings representing a particular "ideal shape" that some score is capturing likewise
will not in general be diagrammable. It will generally not be possible to construct data representing the pure type of one component
without any admixture. The necessary distances will again be incon-
sistent with the geometry of E^2 or E^3 with which we began. In reviews of multivariate approaches to morphometrics, such as Blackith
and Reyment (1971) and Corruccini (1975), there are never any drawings of biological forms; for if the analysis cannot be depicted
except in canonical space, there is no reason to show the data.

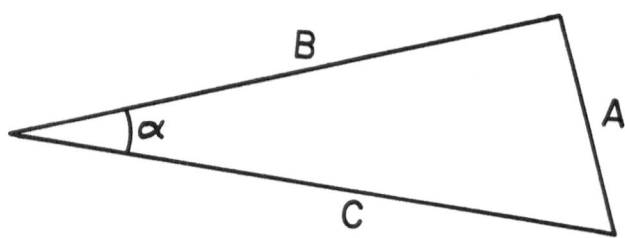

Fig. IV-3.--For a population of triangles like this one, the
means of sides A, B, C delimit a triangle which does not necessarily
have an angle equal to the mean of the angles α.

To be diagrammable, the output of a statistical analysis must
satisfy certain precise constraints deriving from the geometric model
for the measures. There are generally two algebraic forms of the
constraint, one for the 2N-vector means over populations or subpopu-
lations of comparable whole shapes, another for small deviations
therefrom. For the turning-tangent representation I propose here,
the criterion for diagrammability is particularly simple:

$$\oint \exp\,(i\theta(s))ds = 0,$$

where ds is the element of arc-length, $\theta(s)$ is the azimuth there
(the net tangent angle measured from an arbitrary zero), and i is
the square root of -1. In words: an outline is diagrammable whenever the net displacement of a point as it goes around once is precisely zero. I call this the closure criterion.

Let our measurement 2N-vector, one vector per outline, have for
components the arc distances p_i, $0 < p_1 < p_2 < \ldots < p_N < 2\pi$, of N predetermined
landmarks relative to an $N+1^{th}$, the origin, and the tangent angles
θ_j, $j=1, \ldots, N$, at the first N landmarks measured relative to the
angle at the $N+1^{th}$. Define $d_j = .5(p_{j+1} - p_{j-1})$, $j=2, \ldots, N-1$, and
$d_1 = .5(p_2 - (p_N - 2\pi))$, $d_N = .5((p_1 + 2\pi) - p_N)$. Then the integral in the
closure criterion can be approximated by a finite sum:

$$\sum_j \exp(i\theta_j) d_j = 0 .$$

This equation is satisfied approximately by all the real data, and
the greater N is, the more exactly it holds.

Let $\bar{p}_j$, $\bar{\theta}_j$ be the true means over comparable shapes of the p_j,
θ_j respectively. They will in general not satisfy the closure criterion, but for well-behaved data they will not fail by much. Define
pseudomeans $\tilde{p}_j$, $\tilde{\theta}_j$ which best fit the actual p's and θ's subject to
the constraint of exact closure, and write $\tilde{\pi}_j = \tilde{p}_j - \bar{p}_j$, $\tilde{\tau}_j = \tilde{\theta}_j - \bar{\theta}_j$, $\tilde{\delta}_j = $
$.5(\tilde{\pi}_{j+1} - \tilde{\pi}_{j-1})$. Then, by the closure criterion,

$$0 = \sum \exp(i(\bar{\theta}_j + \tilde{\tau}_j))(\bar{d}_j + \tilde{\delta}_j)$$

$$\cong \sum \exp(i\bar{\theta}_j)(1 + i\tilde{\tau}_j)(\bar{d}_j + \tilde{\delta}_j).$$

Assuming $\tilde{\delta}_j \tilde{\tau}_j$ negligible, of second order in small quantities, we
derive the nonhomogeneous complex constraint

$$- \sum \exp(i\bar{\theta}_j)\bar{d}_j = \sum \exp(i\bar{\theta}_j)(\tilde{\delta}_j + i\bar{d}_j\tilde{\tau}_j).$$

For purposes of computation we reduce this to two constraints, equat-
ing real and imaginary parts separately. We substitute the defini-
tion of the $\tilde{\delta}$'s to phrase this in terms of the $\tilde{\pi}$'s. Under this pair
of nonhomogeneous linear constraints we wish to minimize the total
deviation of our pseudomeans from the true values: this total may be
taken as $\sum \bar{d}_j(\tilde{\pi}_j^2 + \tilde{\tau}_j^2)$. Application of two Lagrangian multipliers

reduces this to a straightforward problem in matrix algebra. See
Mikhail (1976: ch.7).

Given the values $\bar{\theta}_j + \tilde{\tau}_j$ for the pseudomeans of θ_j and $\bar{d}_j + .5(\tilde{\pi}_{j+1} - \tilde{\pi}_{j-1})$ for the pseudomeans of the d_j, we replace the variance-covariance matrix of the 2N variables $\{d_j, \theta_j\}$ with crossproducts centered about the pseudomeans. Any coefficient vector is a pattern $\{\hat{d}_j, \hat{\theta}_j\}$ of loadings. Closure demands that $\sum \exp(i(\bar{\theta}_j + \varepsilon\hat{\theta}_j))(\bar{d}_j + \varepsilon\hat{d}_j) = 0$ for small scores ε. This linearizes to constrain whatever matrix computations are involved in component extraction, discriminatory analysis, and the like, by the two homogeneous criteria which are real and imaginary parts of

$$\sum \exp(i\tilde{\theta}_j)(\hat{d}_j + i d_j \hat{\theta}_j) = 0 .$$

Whatever we compute is then automatically diagrammable, perhaps by a
superposition of the three outlines described by $\{\tilde{d}_j, \tilde{\theta}_j\}$ and $\{\tilde{d}_j \pm \hat{d}_j, \tilde{\theta}_j \pm \hat{\theta}_j\}$.

For data restricted to landmark coordinates only, without information about arc-length or tangent angle, I propose a slightly different analysis. The positions of the landmarks relative to their
neighbors--the vectors connecting successive landmarks around the
curve--may be measured as ordinary complex numbers in the plane,
then submitted to the Hermitian multivariate analysis suggested below for the analysis of skeletons, p. 61. Difference scores of successive vectors after a complex logarithmic transformation will
clear the computations of orientation and scale; there will result a
factor analysis of segmented shape in the plane, as defined in
chapter ii--an analysis suited for profiles of cephalic soft tissue
and other complexly twisting forms. Should the data in fact be
closed polygons, a closure constraint must be applied to the multivariate analysis, in a form quite similar to the preceding.

3. Conic replacement curves and their estimation

As a simple function of one variable, the intrinsic representation of an outline by tangent angle as a function of arc-length has
many statistical advantages. It includes information both local and
global; it can be sampled at intervals, or at "landmarks," and put
into a multivariate analysis. It begs no questions of feature significance. Unfortunately, such a function is never given us in the
course of empirical research. We have, rather, selected points upon
such a curve, in the manner of Fig. IV-4.

Fig. IV-4.--Typical data sets for outline-fitting: functional
relationships of unknown structure. Empirical arc-length does not
exist.

Now these isolated points must be combined into a smooth curve
somehow. It is clear we cannot connect the given points with straight
lines, for the resulting function has tangent angle undefined at
exactly the points, now vertices of a polygon, at which we need to
sample it. In case the data are chain-coded, the situation is even
worse, for every azimuth is one of the values 0°, 90°, 180°, 270°.
Nor can we say that the data approximate to a "true" curve, that with
greater and greater density of points sampled we would converge to
a limiting "true" locus. At the limit, tangent angle is not a continuous function of arc-length, for arc-length does not exist. The
greater the detail in which one examines any real locus, natural or
fabricated, the more irregular it appears, the longer the arc-length
and the denser its local extrema of curvature. See Steinhaus (1954),
Mandelbrot (1977) for a discussion of this remarkably general phenomenon, that arc-length generally increases exponentially with the
scale of observation. This paradox is not peculiar to biometrics;

it complicates also the study of seacoasts, galaxies, turbulence, evolutionary histories, and soap.

Then to utilize the intrinsic representation theorem in data analysis it is necessary to construct a rectifiable curve fitting the data before we proceed. We could, for instance, pass a curve suitably smooth through every point of the digitized outline, thus enshrining all measurement error on a par with form itself. Such curves would tend to be a good deal longer than what we intuitively feel is the appropriate "length" of the outline. We could contrariwise try to capture the form by a "best-fitting" specimen of some family of simple mathematical curves. These families, however, would have to have a number of parameters which is some multiple of the count of "pieces" of outline (arcs between landmarks) to be satisfactorily fit, and such families do not come readily to mind. It is best to proceed by splining, which is to say, computing a linked series of curves, each of which best fits the data of its sub-arc of the outline and which pass smoothly each into its neighbors. The individual pieces which we spline together should be conic sections, the next simplest family of plane curves after straight lines.

Fit of a conic to a point. To fit conic splines to closed outlines, it is necessary first to set a reasonable criterion for the fit between a point and a conic curve, so that we can compute that curve from our family whose fit is exactly best. For the usual sort of curve-fitting, which estimates the mathematical form of a dependency function, error from a curve is measured "up" or "down," in the direction of the dependent-variable axis on a scatterplot. In the present enterprise, with no vertical alignment assignable, it is not so clear how one is to proceed.

In the course of investigations into the shape of plane curves, several authors, attempting to fit conic sections to scatters, have recently and independently discovered algorithms that are reducible to conventional linear computations. All minimize the sum-of-squares of a form $Q(x,y)=Ax^2+Bxy+Cy^2+Dx+Ey+F$ over a scatter in the (x,y)-plane. Paton (1970a,b) sets $A^2+B^2+C^2+D^2+E^2+F^2=1$ as the constraint on this minimization, whereupon the solution vector (A, B, C, D, E, F) is a principal component of the scatter matrix of the variables x^2, xy, y^2, x, y, 1. Biggerstaff (1972), Albano (1974), and Cooper and Yalabik (1976) set $F=1$ as the constraint, whereupon there arises a set of normal equations for A, B, C, D, E in terms of the same

scatter matrix. Gnanadesikan (1977) sets $A^2+B^2+C^2+D^2+E^2=1$, yielding
for extremum the last principal component of the variance-covariance
matrix of x^2, xy, y^2, x, y. All these authors loosely characterize
the optimal conic as "best-fitting" of all the conics of the plane,
but the nature of the fit each is optimizing, whose computation is
so surprisingly straightforward, has not been elucidated.

We seek an estimation rule which is <u>general</u>, <u>simple to compute</u>,
and <u>invariant</u>. Generality requires a single algorithm by which we
fit both ellipses and hyperbolas, with circles and parabolas included
as numerically special cases of probability zero. For instance, we
cannot generalize the fitting of parabolas usual in quadratic regression, in which error-of-fit is measured parallel to the axis of the
parabola, because in the general conic there are two axes, two directions through every point parallel to the axes, and upon each direction zero, one, or two points of intersection with the conic.

Simplicity suggests that we eschew "non-linear," iterative techniques whenever possible, and instead settle upon algorithms whose
estimates can be computed from the data by finite arithmetic formulas or by the standard inversions and eigenextractions of matrix
calculus. The solution I shall propose is computable so; several
alternatives, such as the minimization in mean square of distance
perpendicular to the conic, involve expressions in roots of bilinear
combinations in the data and the unknowns, and cannot be optimized
in closed algebraic form.

Invariance is to be with respect to the equiform group of
transformations of the Euclidean plane: rotations, translations,
and changes of scale. If the coordinates of the data be transformed,
point by point, according to an element of this group, then the best-
fitting conic to the new scatter must be exactly the result of the
same transformation applied to the conic which was best-fitting to
the original scatter. The estimation rule I shall propose minimizes
a quadratic form in the unknown conic parameters subject to an exact
constraint on the value of another quadratic form. I will show that
both of these forms are invariants with respect to the equiform
group: one is proportional to a sum of squared distance-ratios--the
affine group preserves these, and the equiform group is a subgroup;
the other is a linear combination of other forms generally known to
be invariant for conics under the action of the equiform group.
Invariance of minimand and constraint together implies the invariance
of the net estimation procedure.

Geometric interpretation. Let a given central conic section
have equation $Q(x,y)=Ax^2+Bxy+Cy^2+Dx+Ey+F=0$. Let (x_0,y_0) be an arbitrary point of the plane, and set (x_1,y_1) equal to the point on
the conic on the ray from the center through (x_0,y_0).

Choose a new coordinate system, denoted by primes, which is centered at the center of the conic and aligned with its principal axes.
Define the form $Q'(x',y')$ on the new coordinates so that $Q'(x',y')=$
$Q(x,y)$ identically on the plane (x',y'). In this new coordinate system, $Q'(x',y')=A'x'^2+B'y'^2+F'$ for some A', B', F'.

In the primed system, the center is the origin, $(0',0')$. By
collinearity of (x_0,y_0), (x_1,y_1), and the center, expressed in the
primed coordinate system, we have $x_1'/y_1'=x_0'/y_0'$ or $x_0'=x_1'(y_0'/y_1')$. Then

$$Q(x_0,y_0)=Q'(x_0',y_0')=A'x_0'^2+B'y_0'^2+F'$$

$$=(y_0'/y_1')^2 (A'x_1'^2+B'y_1'^2)+F'$$

$$= (y_0'/y_1')^2(-F')+F'$$

$$= -F'((y_0'/y_1')^2 - 1), \tag{1}$$

since (x_1',y_1') is on the conic $Q'(x',y')=0$. But the term (y_0'/y_1')
is just the ratio of distances center-to-point and center-to-conic
along the ray we are using--the ratio of $d+d_1$ to d in Fig. IV-5; for
projection onto any axis preserves distance along a line.

Hence $Q(x_0,y_0) \propto ((d+d_1)^2/d^2)-1$, and as the equiform transformations do not alter distance ratios they preserve this proportionality
(though they may alter F').

We have $((d+d_1)^2-d^2)/d^2=d_1(d_1+2d)/d^2$. When the data are quite
close to a conic, d_1 is small, d_1+2d is approximately equal to $2d$,
and $Q(x_0,y_0)$ is approximately proportional to d_1/d, the distance
from the conic along a line through the conic's center, measured in
units of distance from the center to the conic in that direction.

The fitting of a parabola is a limiting case, exactly transi-
tional between ellipse and hyperbola. As the center of an ellipse
moves off toward infinity while its major axis and the curvature of
one end are held constant, the rays from its center become parallel,
and the difference of squared distances which is the numerator of Q
becomes linear in distance. The quantity being minimized in the

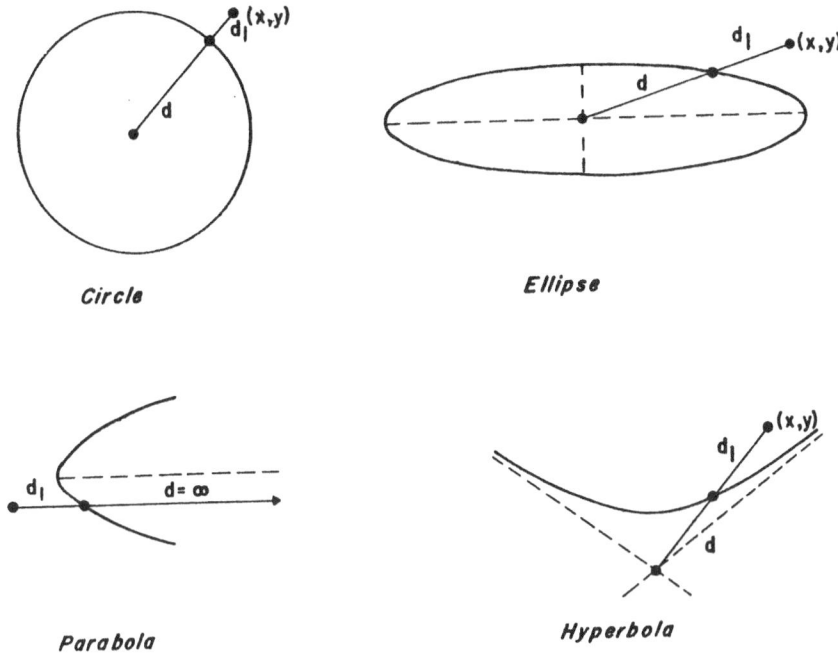

Fig. IV-5.--The geometry of "error-of-fit" about a conic section as measured by the equation of the conic. The error is always
proportional to $((d+d_1)/d)^2 - 1$.

fitting then approaches the usual deviation, Euclidean distance
from the curve measured parallel to the axis.

Estimator for a circle. For a circle, I propose using the instance of least mean-squared Q. Let there be given points (x_1,y_1),
..., (x_n, y_n) and let the sample linear regression of x^2+y^2 on x, y,
1 be $x^2+y^2 = 2ax + 2by +c$, where a, b, and c are the least-squares
estimates. Then the circle centered at (a,b) of radius $(c+a^2+b^2)^{1/2}$
is best-fitting when error is measured by Q(x,y) as defined in the
previous section.

[Proof: for a circle, $Q(x,y)=(x-a)^2+(y-b)^2- R^2 = x^2 + y^2 - 2ax$
$- 2by - (R^2 - a^2 - b^2)$, where R is its radius. As the relation

between the parameters (a, b, R) of Q and the parameters (2a, 2b, R^2 - a^2 - b^2) of the regression is one-to-one, we can minimize error sum-of-squares with respect to one set of parameters by minimizing . it with respect to the other. But minimization with respect to the second set corresponds to the ordinary least-squares regression of x^2+y^2 on x, y, 1, as asserted.]

There is a reasonable probabilistic model which yields this least-squares estimate as the maximum-likelihood solution for its parameters. Let data points (x_i,y_i), i=1, ..., n, be scattered around a circle of center (x_0,y_0) and radius R. Let (r,θ) be polar coordinates about (x_0,y_0) of any one of these data points. (Then $r^2 = (x-x_0)^2 + (y-y_0)^2$.)

Let the probability element for the data scatter be

$$dF = C \frac{1}{\sigma\sqrt{2\pi}} \exp(-(r^2 - R^2)^2 / 2\sigma^2)d(r^2)d\theta/2\pi$$

--truncated normal in r^2 about R^2, uniform in θ. (For σ small with respect to R, this will look just like a normal distribution in r about R.) The value of C is $(1 - \Phi(-R^2/\sigma))^{-1}$, where Φ is the standard cumulative normal function.

The Jacobian of the transformation from (r^2,θ) coordinates to Cartesian coordinates (x,y) is the constant 2. That is to say, $d(r^2)d\theta=2r\ dr\ d\theta=2\ dx\ dy$. Then

$$dF = C/\pi \frac{1}{\sigma\sqrt{2\pi}} \exp(-((x-x_0)^2+(y-y_0)^2 - R^2)^2 / 2\sigma^2)\ dx\ dy.$$

Purely as a formal device, we set quite flat improper prior probabilities on (x_0,y_0,R,σ) as follows:
the prior of (x_0,y_0) is absolutely flat;
the prior of (R,σ) is independent of (x_0,y_0) and equal to C^{-n}. It ranges between $1/2^n$ and 1, and is large when σ is small compared with R^2.

Then for this probabilistic model with this nearly flat prior, the posterior probability of the parameter set (x_0,y_0,R,σ) given the data is

$$\frac{1}{(\sigma\sqrt{2\pi})^n} \exp\left(- \sum_1^n ((x_i-x_0)^2 + (y_i-y_0)^2 - R^2)^2 / 2\sigma^2 \right).$$

The parameters for maximum posterior probability are those for which

$$- n \log \sigma - \sum_{1}^{n} ((x_i-x_0)^2 + (y_i-y_0)^2 - R^2)^2 / 2\sigma^2$$

is maximal; by the usual calculations appropriate to maximum likeli-
hood the maximum can be located at the minimum of the summation; but
this is just the circle of least mean-squared Q, whose computation is
equivalent, I showed above, to an ordinary regression.

Since Q is an expression in distances $r^2 - R^2$, it is invariant
under rotation and translation. Under change of scale (say, multiplication by k about a fixed origin), x^2+y^2 is multiplied by k^2, x,
y each by k. Then the coefficients of the regression are multiplied
by k, except the constant, which is multiplied by k^2. The best-
fitting circle thus has coordinates of its center multiplied by k,
and its squared radius multiplied by k^2; therefore the entire fitting procedure is invariant under equiform transformations, as required.

In the case that data arise from an arc of a circle of extent
2α considerably less than 2π radians, the preceding computation is
unchanged in all essentials. The element of arc is now $d\theta/2\alpha$ instead of $d\theta/2\pi$, and the constant C is replaced by a new constant C'
$=C/\alpha$. Now α is a function of R, though not of σ; for σ small, R ·
sin α = d/2, where d is the distance from one "end" of the data to
the other. Then C' = C/α $\propto$ $1/\sin^{-1}(d/2R)$, and our prior on (R,σ)
must be multiplied by $(\sin^{-1}(d/2R))^n \propto R^{-n}$ for α small. We are in
effect biassing ourselves against circles of larger radius.

Estimation for the general conic. For an arbitrary conic, the
generalization of the error-of-fit $(x-x_0)^2 + (y-y_0)^2 - R^2$ is clearly
the value $Q(x,y) = Ax^2 + Bxy + Cy^2 + Dx + Ey + F$. We need a normalization of these coefficients, for the conic described by $Q(x,y)=0$ is
the same as that described by $kQ = kAx^2 + \ldots + kF = o$ for any k,
though the latter has error sum-of-squares k^2 times that of Q, and so
can be as small as we wish. The norm we seek should be a quadratic
form in A, B, C, D, E, F, for simplicity of computation. It must be
positive-definite, or else certain conics, those for which the form
happens to be exactly 0, could never be fitted at all, though the
data lay exactly upon them. It cannot involve D, E, or F, since these
are functions of location of the origin of coordinates and become arbitrarily huge when the center is translated. The norm must therefore be a function of A, B, C. Now in the case of lines (cf. Pearson,

1901), the line which is best-fitted in the sense of distance perpendicular to itself can be estimated invariantly by minimizing $\Sigma_{data}(ax+by+c)^2$ subject to the constraint $a^2+b^2=1$. We might try, by direct mimicry, the form $A^2+B^2+C^2$. Unfortunately, the hyperbola $2xy = 1$, with $A^2+B^2+C^2 = 4$, can be rotated into the form $x^2 - y^2 = 1$, with $A^2+B^2+C^2 = 2$, and so the first would automatically fit twice as badly if the data were rotated $45°$; but the selected normalization must be invariant under rotation. Introductions to the algebraic theory of conics tell us that the forms $A + C$ and $B^2 - 4AC$ are invariant under the Euclidean group. The only positive-definite invariant that can be formed from these quantities is $(A + C)^2 +$ $(B^2 - 4AC)/2 = A^2 + B^2/2 + C^2$. This is never zero for the non-degenerate conics, and so we may set a unique scale for 0, and thus proceed to actual minimization of ΣQ^2, by setting this norm equal to some conventional constant value. I suggest the value 2, so that the equation of a circle is in the usual form.

Given data points (x_1,y_1), ..., (x_n,y_n) scattered about a conic, I propose to estimate the conic of best fit by minimizing $\Sigma_i (Q(x_i,y_i)^2)$ subject to the constraint $VDV' = 2$, where $V =$ (A,B,C,D,E,F) is the vector of coefficients to be estimated, and D is the matrix diag $(1,1/2,1,0,0,0)$. The minimand may be written VSV', where S is the scatter matrix about 0 of the 6-vector $(x_i^2,$ x_iy_i, y_i^2, x_i, y_i, 1) as a function of the data points (x_i,y_i).

If the coordinates of the data points be rotated or translated, the constraint $A^2 + B^2/2 + C^2$ is unchanged, likewise the term F' of equation (1), which is gotten by undoing the transformation; and the distance ratio, the other multiplicand of (1), is clearly unchanged. Then the fitting is invariant under translation and rotation. If the coordinates be rescaled by k, $(x,y) \rightarrow (x',y') = (kx,ky)$, let V_k be the vector of coefficients of the resulting best-fitting conic by this algorithm. The conic $Ax^2 + ... + F = 0$ becomes, after rescaling, $A(x'^2/k^2) + ... + F = 0$ or, renormalizing to the same constraint, $Ax'^2 + Bx'y' + Cy'^2 + kDx' + kEy' + k^2F = 0$. The rescaling has replaced the matrix S by $S_k = D_k SD_k$ with $D_k = $ diag $(k^2,k^2,k^2,k,k,1)$. Note that $D_k DD_k = k^4D$. Extremization of $V_k(D_k SD_k)V_k'$ for $V_k DV_k$ constant is equivalent to extremizing it for $V_k D_k DD_k V_k'$ constant. Since D_k is nonsingular, the extremum is V/D_k, where V is the extremum (A, B,C,D,E,F) before rescaling. But V/D_k gives the coefficients of the old conic in the new system, for $k^2V/D_k = (A,B,C,kD,kE, k^2F)$. Hence this method of fitting is invariant under equiform transformations.

As the authors I cited on pages 36-37 all use D and E in the norms for their fitting, none of their solutions are invariant under the Euclidean group. Those which involve F, further, are not even invariant under translation; for by setting F=1 it becomes impossible to fit conics through the origin (F=0) at all. These authors must therefore forcibly standardize their data before conic fitting in any of various ad hoc ways. Translation of center-of-mass and rotation to principal axes of scatter are the most common. In my opinion, this will not do. Data must be fit in a manner invariant under the Euclidean group, not arbitrarily constrained with respect to it.

Computation of the extremum. We wish to minimize VSV' subject to VDV' = constant. Partition V into $(V_1|V_2)$, both components of length 3, and let S be partitioned correspondingly:

$$S = \begin{pmatrix} S_{11} & S_{12} \\ S_{21} & S_{22} \end{pmatrix}.$$

Then

$$VSV' = V_1 S_{11} V_1' + 2 V_1 S_{12} V_2' + V_2 S_{22} V_2'.$$

We must minimize this subject to $V_1 D_1 V_1'$ = constant, where D_1 = diag (1, 1/2, 1). For any fixed V_1, VSV' is minimal when

$$d(VSV')/dV_2 = 2V_1 S_{12} + 2V_2 S_{22} = 0$$

which implies

$$V_2 = -V_1 S_{12} S_{22}^{-1}.$$

Then

$$VSV' = V_1 (S_{11} - S_{12} S_{22}^{-1} S_{21}) V_1' = V_1 S_{11 \cdot 2} V_1'.$$

To minimize this for $V_1 D_1 V_1'$ = constant, let λ be a Lagrangian multiplier for the constraint. Then we must set to 0 the derivative with respect to V_1 of $V_1 S_{11 \cdot 2} V_1' - \lambda V_1 D_1 V_1'$. This yields

$$2 S_{11 \cdot 2} V_1' - 2\lambda D_1 V_1' = 0,$$

so that λ is a relative eigenvalue of $S_{11 \cdot 2}$ with respect to D_1--that is to say, a solution of $|S_{11 \cdot 2} - \lambda D_1| = 0$ --and V_1 is the corresponding eigenvector. The eigenvector of best geometrical fit is the one we want, usually (but not always) that of smallest λ.

The matrix $S_{11 \cdot 2}$ is the covariance matrix of the residuals of x^2, xy, y^2 after regression on x, y, 1; and the vector $V = (V_1|V_2)$ $= V_1(I|-S_{12}S_{22}^{-1})$ corresponds to residualization of the basis of the V_1-space, e.g. subtraction of the part of each of x^2, xy, y^2 which is linearly predicted by x, y, 1. Geometrically, in the space of V we find the subspace completely orthogonal to the null space of D-- this is the subspace of residuals of the second-order terms on those of lower order--and perform our optimization within that subspace.

Linear constraints. This computation remains tractable when we place arbitrary linear constraints upon the parameters. If V be used to denote the 6-vector of coefficients we are estimating, and a set of constraints be written together in the form VM = 0, M a matrix, then Rao (1973: sec. 1c.6) instructs us in this formula for the extremum we seek: the vector V of smallest VSV' subject to the constraints VDV' = constant, VM = 0 is the eigenvector of largest eigenvalue of the matrix $(I - M(M'S^{-1}M)^-M'S^{-1})D$ with respect to S. Here - denotes any generalized inverse. The matrix S must be positive definite, e.g. have an ordinary inverse, whereas D may be singular and the constraints may be redundant.

Setting linear constraints erodes the available degrees of freedom in various ways, some quite useful. Here are some possibilities:

(1) B = 0. The conic's principal axes are parallel to the coordinate axes.

(2) A = C. The conic's principal axes are parallel to y = ±x.

(3) A = C, B = 0 (two constraints). The conic is a circle.

(4) B = 0, C = 0. The conic is a parabola with axis vertical.

(5) $2Ax_0 + By_0 + D = 0$, $Bx_0 + 2Cy_0 + E = 0$. The conic has its center at (x_0, y_0).

(6) $Ax_1^2 + Bx_1y_1 + Cy_1^2 + Dx_1 + Ey_1 + F = 0$. The conic goes through (x_1, y_1).

(7) $(2Ax_1 + By_1 + D) \cos \alpha + (Bx_1 + 2Cy_1 + E) \sin \alpha = 0$. If the point (x_1, y_1) is on the conic, this is the criterion that the tangent at (x_1, y_1) make angle α with the positive x-axis.

(8) $A (- x_1 \sin \alpha + R \cos^2 \alpha) + B (- .5y_1\sin^2 \alpha + .5x_1\cos^2 \alpha + R \cos \alpha \sin \alpha) + C (y_1 \cos \alpha + R \sin^2 \alpha) + D (- .5 \sin \alpha) + E (.5 \cos \alpha) = 0$. The conic through (x_1, y_1) with tangent making

an angle α with the positive x-axis is by this constraint required to
have radius of curvature there equal to R.

4. Conic splining

There is a rapidly growing literature (cf. Poirier, 1973; Wold,
1974) on polynomial spline regression. In this technique, the expected value of a dependent variable is fitted by different func-
tions--usually polynomials--over distinct ranges of an independent
variable. At "knots," where these ranges abut, the predictions are
required to agree in value and in their first few derivatives. After the knots are set, all the polynomials are computed at once, by
minimizing the total squared error-of-fit about all of them in one
clever dummy-variable regression. The method becomes much more
complex when error-of-fit is not always measured along a fixed axis.
The constraints which force the curves to line up become nonlinear
in the coefficients being estimated, and linear methods fail (cf.
Thomas, 1976, and references therein). In particular, one cannot
spline a closed curve linearly by polynomials. But just as parabolas (and higher-order polynomial graphs) generalize straight-line
regression, so do conics (and higher-order algebraic curves) generalize straight-line curve-fitting in general orientation. It is then
of interest to explore the possibility of fitting multiple conics to
separate parts of a curve simultaneously in a single linear computation generalizing the previous discussion for fitting conics singly.

Joint conic fitting under constraint. Suppose we have a data
set in the plane which looks like two conics, in the manner of
Fig. IV-6, impinging upon each other at some visible nexus P. It
would be nice to fit two conic segments simultaneously, one to the
left of P, one to the right, which minimize the net error-of-fit.
We would like the conics to connect at P without any corner, that
is, to have the same tangent there. The inclination of this mutual
tangent is not given in the data, however; we must estimate it, too,
as that position of the tangent line at P for which the best two
conics through P with that tangent, one for each side, together have
least mean-squared error-of-fit. To execute this plan we must link
the normalizations of the two conics. The best way to do this is
by insisting that a given error-of-fit near P be the same whichever
conic we consider it on. To a linear approximation, the error for
a point such as R relative to a conic $Q(x,y) = 0$ through P is equal
to the gradient of the scalar field $Q(x,y)$ times the distance from

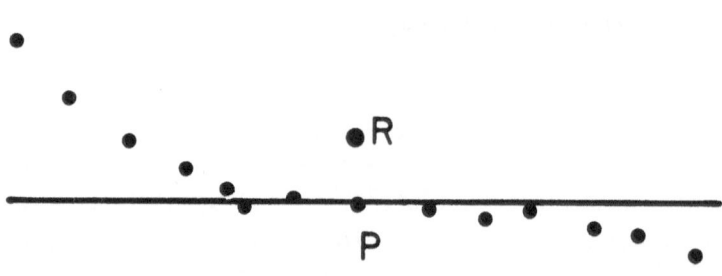

Fig. IV-6.--Two conics abutting at a knot where the tangent
crosses the curve.

P to R normal to the curve. These will be equal for the two conics
impinging at P if and only if the gradients of the two scalar fields
are equal at P.

But to specify that these two gradients be equal is just a pair
of linear constraints, for the gradients are linear forms in the coefficients of the conics. To wit: if the two conics be $Q_i(x,y) =$
$A_i x^2 + B_i xy + C_i y^2 + D_i x + E_i y + F = 0$, $i = 1$, 2, and if P have the
coordinates (r,s), then the two consistency criteria are

$$2A_1 r + B_1 s + D_1 - (2A_2 r + B_2 s + D_2) = 0,$$

$$2C_1 s + B_1 r + E_1 - (2C_2 s + B_2 r + E_2) = 0.$$

We must also insist that both conics actually pass through the point
(r,s) at which we are specifying their tangents, and this is two more
constraints:

$$A_i r^2 + B_i rs + C_i s^2 + D_i r + E_i s + F_i = 0, \quad i = 1, 2.$$

We can estimate both conics, all twelve coefficients, simultaneously
by minimizing their total squared error-of-fit subject to the normalization $A_1^2 + B_1^2/2 + C_1^2 = 2$ and the four constraints preceding.

This tactic is easily generalized for the fitting of conic
splines to extended chains, even closed curves. Let there be I arcs,
and on the ith arc data points (x_{ij}, y_{ij}), j=1, ..., n_i, i=1, ..., I.

Let the arcs abut at knots (x_i, y_i), i=1, ..., I, and set (x_{I+1}, y_{I+1})
= (x_1, y_1). Let S_i be the 6 x 6 scatter matrix about 0 for the points
$(x_{ij}^2, x_{ij}y_{ij}, y_{ij}^2, x_{ij}, y_{ij}, 1)$ derived from the data points $(x_{ij},$
$y_{ij})$ of the ith arc. Define S, (6I) x (6I), equal to blockdiag
$(S_1, ..., S_I)$. Consider a conic spline which is equal to $A_i x^2 + B_i xy$
$+C_i y^2 + D_i x + E_i y + F_i$ on the ith arc, i=1, ..., I, and let $A_{I+1} = A_1, ...,$
$F_{I+1} = F_1$. The error sum-of-squares about the ith arc for the data
of the ith arc is $(A_i, B_i, C_i, D_i, E_i, F_i)S_i(A_i, B_i, C_i, D_i, E_i, F_i)'$. Define
$V = (A_1, ..., F_1, A_2, ..., F_{I-1}, A_I, ..., F_I)$, a (6I)-vector. Then
the total sum-of-squares for error-of-fit about all the arcs is VSV'.
This may be minimized subject to a set of homogeneous linear constraints VM=0 by the formula of Rao quoted above. In this application, S must be nonsingular; then each S_i must be nonsingular, which
implies that $n_i \geq 6$, all i, and that the points of all arcs must lie
in "general position" (not all be upon one conic).

We may minimize VSV' subject to up to 4I constraints of inci-
dence and tangency:

$$A_i x_j^2 + B_i x_j y_j + C_i y_j^2 + D_i x_j + E_i y_j + F_i = 0, \quad i=1, ..., I, \ j=i, \ i+1;$$

$$2A_k x_k + B_k y_k + D_k - (2A_{k-1}x_k + B_{k-1}y_k + D_{k-1}) = 0, \quad k=1, ..., I;$$

$$2C_k y_k + B_k x_k + E_k - (2C_{k-1}y_k + B_{k-1}y_k + E_{k-1}) = 0, \quad k=1, ..., I;$$

and also subject to a normalization. As the tangency constraints explicitly link normalizations of one conic to the next, we can only
assign one normalizing constraint for any subarc chain of continuously
turning tangent. In the usual case where this subarc is the whole
spline, as in Figs. IV-7 through IV-10, this means that any constraint
$A_i^2 + B_i^2/2 + C_i^2$ = constant determines the normalization of all the arcs
all the way around.

An example. The vaguely biological outline, Fig. IV-7(top),
de rives from a free-hand sketch on graph paper. I chose three
"landmarks": #1, the "corner" in lower center; #2, the extreme
rightmost point; and #3, the extreme topmost point. If this were a
skull x-ray, #1 might be the auditory meatus, #2 the bridge of the
nose, and #3 the bregma. Points in between the landmarks have been
digitized crudely by eye.

The spline arcs are computed by a computer program executing the
following algorithm.

   a. The data are read in arc by arc and the scatter matrix S

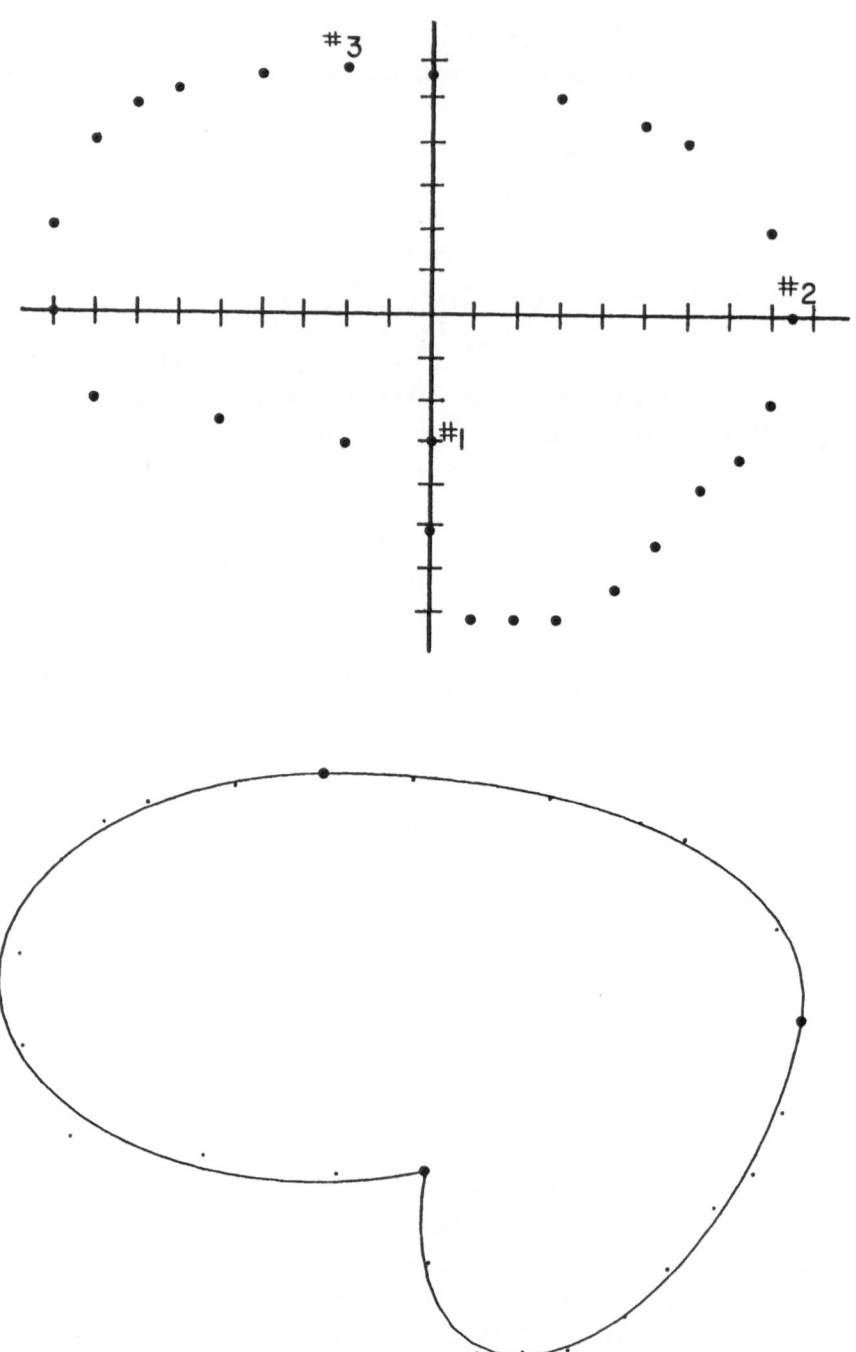

Fig. IV-7.--(Top) Hypothetical data set for conic splining.
(Bottom) A three-arc conic spline approximating this scatter, with
knots at the three large points.

accumulated and stored. A vector of 18 unknown coefficients $V = (A_1,$ ..., F_1, ..., A_3, ..., F_3) is allocated, bearing the conic coefficients arc-wise.

b. The constraints are written out in full. Arcs 3 and 1 pass through the point #1 of coordinates (0, -3). This requires

$$A_i \cdot 0 + B_i \cdot 0 + C_i \cdot 9 + D_i \cdot 0 + E_i \cdot (-3) + F_i = 0, \quad i=3, 1, \text{ or}$$

$$9C_3 - 3E_3 + F_3 = 0, \tag{1}$$

$$9C_1 - 3E_1 + F_1 = 0. \tag{2}$$

Similarly, arcs 1 and 2 pass through the point #2 = (8.5,0), requiring

$$72.25A_1 + 8.5D_1 + F_1 = 0, \tag{3}$$

$$72.25A_2 + 8.5D_2 + F_2 = 0; \tag{4}$$

and for the point #3 = (-2,5.7),

$$4A_2 - 11.2B_2 + 32.49C_2 - 2D_2 + 5.7E_2 + F_2 = 0, \tag{5}$$

$$4A_3 - 11.4B_3 + 32.49C_3 - 2D_3 + 5.7E_3 + F_3 = 0. \tag{6}$$

At landmark #1 we allow the arcs to intersect at arbitrary angle, and there are no further constraints to be had here. At landmark #2, we pick up a 7th constraint and an 8th from componentwise equality of the gradients of the first conic and the second:

$$17A_1 + D_1 - 17A_2 - D_2 = 0, \tag{7}$$

$$8.5B_1 + E_1 - 8.5B_2 - E_2 = 0. \tag{8}$$

The final two constraints derive from the same reasoning at point #3:

$$-4A_2 + 5.7B_2 + D_2 + 4A_3 - 5.7B_3 - D_3 = 0 \tag{9}$$

$$11.4C_2 - 2B_2 + E_2 - 11.4C_3 + 2B_3 - E_3 = 0 \tag{10}$$

c. For any vector V of coefficients for any spline, the total

error-of-fit is the scalar VSV', where S is the cross-product matrix
accumulated in (a). We need to minimize this for fixed invariant
norm $A_1^2 + B_1^2/2 + C_1^2$ = VDV', D = diag(1, 1/2, 1, 0, 0, ..., 0), subject
to the constraints (1) through (10). Conversely, we can instead maxi-
mize VDV' for fixed error VSV' subject to the same ten constraints.
The theorem of Rao quoted previously applies verbatim here to provide
a computing formula for the solution we seek. A single eigenvector
of a certain large matrix manifests the coefficients of all three
conics, in blocks of six, for its loadings. It is extracted by the
subroutines of the software system "Eispack," as described in Wilkinson and Reinsch (1971). The spline corresponding to this eigenvector
exactly passes through all the knots with well-defined tangents there
and minimizes the total error-of-fit from the conics in between.
This curve is conic between every pair of consecutive landmarks, and
its tangent turns continuously throughout, except where we have expressly omitted the constraint that it do so.

By copying the coefficients of the eigenvector back into the
conic formula, we obtain the equations of our arcs:

- .0160204x^2 - .0008269xy - .0438765y^2 - .0680501x + 1.09105y
  + .722203 = 0 between (0,-3) and (8.5,0);
- .0248356x^2 + .0197886xy - .0153115y^2 + .240555x - .129383y
  - .250344 = 0 between (8.5,0) and (-2,5.7);
- .007336x^2 + .0033159xy - .034512y^2 - .0569287x + .0106358y
  + 1.01396 = 0 between (-2,5.7) and (0,-3).

A check shows the constraints are fit with a maximum error of
1.2×10^{-6}.

d. These arcs are drawn out between appropriate endpoints (by
short segmental approximations) and then superimposed over the data
in Fig. IV-7 (bottom). The spline fits the data quite well except
on the far left, where the conic arc has but one curvature maximum
whereas the data apparently have two (the northwest and southwest
"corners"). This systematic deviation can, if we wish, simply be
declared a "local feature" of the outline relative to the spline (as
the nose is a local feature relative to the plane of the face); it is
not necessarily a flaw in the fitting. The splining-with-landmarks
replaces the "observed" (digitized) boundary with smooth conventional
arcs. Its tangent and normal, turning smoothly, provide a moving
coordinate system to relate deviations from the "standard" in different regions all around the curve. The spline is more elegant and
more easily read than a detailed tracing, as local features do not

disrupt the global aspects of the shape, the ebb and flow of curvature around the form.

Should a landmark, say, the second, be itself known only with error, one can free the conics from passing exactly through it. The revised equations would require, instead of (3), (4), (7), (8), only that the landmark have a common distance from the two conics and the same polar with respect to them. The appropriate constraint equation replaces (3), (4) with the equality of their left-hand sides: $72.25A_1 + 8.5D_1 + F_1 - 72.25A_2 - 8.5D_2 - F_2 = 0$; equations (7), (8) persist unchanged. With this revision, the landmarks serve only a housekeeping function as knots, and do not serve any special role as data themselves.

Application. The technique of conic splining supplies a substitute curve, easily fitted near to data, with a curvature computable simply from the coefficients of the arcs and arc-length equal to a familiar elliptic integral offered in the various collections of scientific subroutines. Regular local features of shape--indentation inside the spline or bumps outside--become new shapes which can be analyzed further. The spline captures the "general shape," and provides an arc-length argument for the local features, in a wholly coordinate-free way. Measures of this general shape can then be derived from inspection of the splined intrinsic function and its variation over populations of shapes, without any ratios, distances, or other "shape variables" having been specified in advance.

Figures IV-8 through IV-10 are examples of conic splines fit to actual biological data. The first of these is a sea-cucumber outline that I digitized from a 3 cm by 5 cm image in an old zoology text. It can be seen how very closely a three-arc conic can be made to follow this creature of very odd shape, a shape, in fact, describable only by reference to this conic spline. The next figure is of a clam shell. The original image here was 6 cm by 6 cm. The fit of two elliptical arcs to this natural form is amazingly close, close enough that the intrinsic curvature function should clearly differentiate among serial stages of this organism's growth and among congeneric adult organisms, by shell outline alone, should anyone care to digitize more of these images.

Figure IV-10 is of more general interest. Of particular irksomeness in craniometrics is the absence of identifiable landmarks over the vault of the human skull. In describing the form of this part of the skull, the only quantities available are distances, angles,

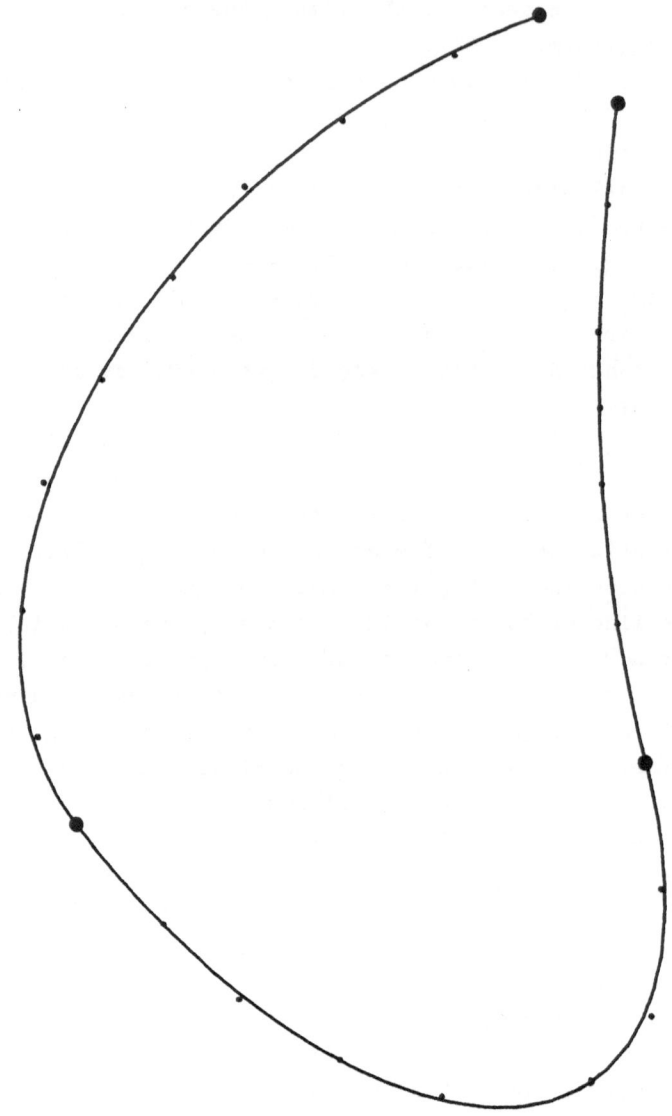

Fig. IV-8.--Three-arc conic spline to a digitization of the outline of a sea cucumber, based on a drawing by Storer (1951:412). The
open area at upper right is the mouth, delimited by fronds and un-
suitable for splining.

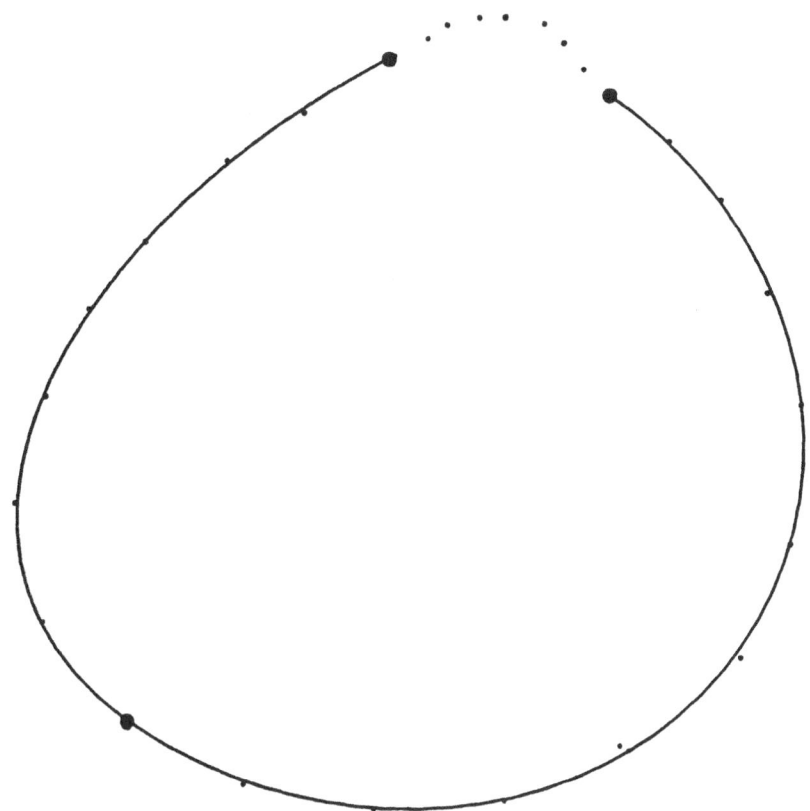

Fig. IV-9.--Two-arc conic spline to digitization of the outline
of a pelecypod, <u>Pisidium</u> <u>compressum</u>, from Burch (1975:14). The open
area at the top is a projection of the hinge obscuring the line of
closure.

and ratios of rather large reach. Figure IV-10 shows a specimen of
this form quite closely fit by either two or three conic arcs splined
together--the optional third arc makes it possible to represent the
little bump above the inion, at the far left. There is then no need
for landmarks at all in this part of the skull, but the whole shape
itself, as smoothly reconstructed by the conic spline, may be used
as a function-valued measure in analysis of growth patterns, racial
differences, evolution of the hominid form, and the like.

   <u>Analysis of parameters</u>. For a sample of these constructions
which closely fit empirical data, one naturally seeks a multivariate

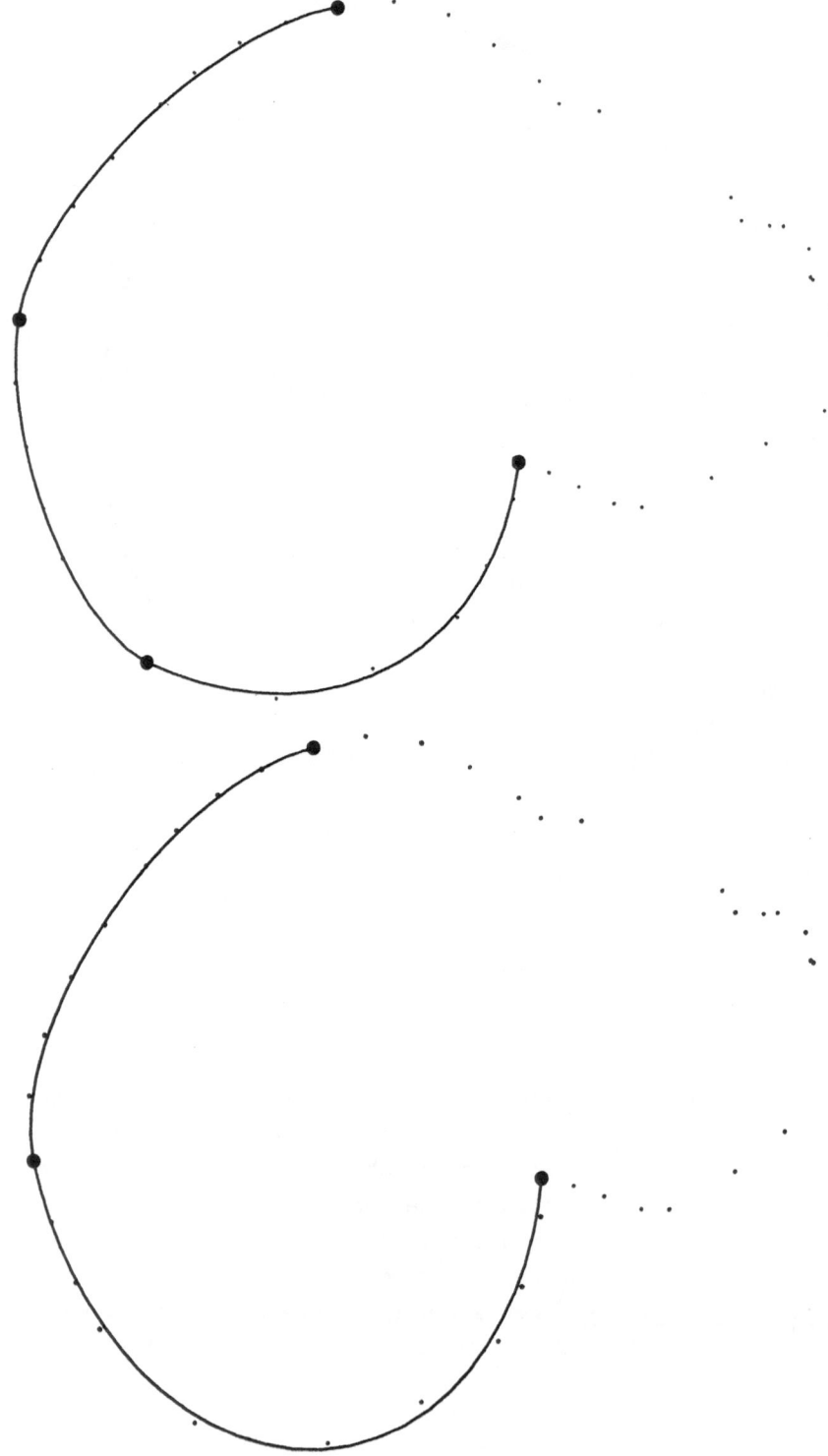

Fig. IV-10.--Two arc and three-arc splines to the calvarium of a sixteen-year-old girl. The data points were selected from the digitization of a cephalogram (from the files of the Biometrics Laboratory of the University of Michigan) constructed by the procedures of Walker and Kowalski (1971, 1972).

analysis of the resulting parameter (6I)-vectors V. We may proceed
by greater or lesser modifications of Bingham's (1974) method for
statistical analysis on the projective plane. The estimated conic
spline is a point in projective space of 6I-1 dimensions, within a
linear subspace, in fact, of co-dimension equal to the number of
linear constraints invoked during the fitting. Let these vectors be
normalized to unit length, i.e., let their components be direction
cosines in the appropriate subspace. By analogy with Bingham's results, which derive from the work of Dimroth and Watson in mathematical geology, the mean is reasonably defined as the first eigenvector of the cross-product matrix of direction cosines, and subsequent eigenvectors represent principal factors of population
variation, pencils of splines about the mean. For single-arc fitting
(I=1) with two constraints, e.g. arcs standardized for orientation
and scale to begin at (0,0) and end at (1,0), there result actual
confidence regions in the picture plane, quartic gores inside which
pre-set fractions of a sample of arcs may reasonably be expected to
lie. These and other matters are investigated in the forthcoming
dissertation of Paul D. Sampson.

B. Extension to Three Dimensions: A Sketch

By viewing the intrinsic function of a curve as a measure
capable of variation, we have uncovered a new methodology firmly
based in the essential geometry of E^2, namely, the exact reconstruction of a curve from its curvature. The mechanism by which we bring
the geometric theory into contact with our data is the device of a
substitute curve.

This notion of substitution can serve as well in three dimensions as in two. To estimate the curvature of a surface-like scatter
at a particular point, we fit the data around that point to a quadric: given data points (x,y,z), we minimize the following equiform
invariant:

$$\sum_{\text{data}} (A_1 x^2 + A_2 y^2 + A_3 z^2 + B_1 xy + B_2 xz + B_3 yz + C_1 x + C_2 y + C_3 z + D)^2$$

subject to the constraint $A_1^2 + A_2^2 + A_3^2 + .5(B_1^2 + B_2^2 + B_3^2) = $ constant.
From the best-fitting equation, by the ordinary reduction to princi-
pal axes, we can visualize a best local shape from the quadric family. We can tell, within statistical tolerance, what the signs and
magnitudes of the principal curvatures are--whether we are looking

at an ellipsoid, an elliptic cylinder, a hyperboloid, hyperbolic
paraboloid, or a plane.

We can go further, at least for blobby, nearly-convex objects,
and spline their whole surfaces in smoothed form, though not by
means of quadrics, and with no closure criterion to enliven the
multivariate analysis. Suppose we have data in the form of successive slices of an outline, a common way of sectioning organisms
either with a knife or with a tomogram. The slices begin at a point
of first contact, a "pole," and widen into lumpy closed curves.
These pass some maximum of sectional area and finally shrink to a
point of final contact at the opposite pole. Let us assume that
every section is star-convex to its intersection with the straight
line connecting the poles. Then the surface is roughly marked out
in lumpy cylindrical coordinates (polar coordinates centered upon
the axis in every section).

By the technique of conic splining, or exact data collection,
let there be available the intersection of a set of meridians with
the bounding curve of the shape in each section. Each such digital
meridian has upon it as many points, as many samples of a radius vector, as there are slices in the data; the first and last radius vectors of a meridian are always zero and always have unbounded derivative with respect to axial position. This poses a problem in the fitting of a spline. We remedy the defect by adapting a device from
potential theory, subtracting off the "principal part" of the singularities at either end. From the best fits of the second and the
penultimate slices to ellipses, we can put an ellipsoid of vertical
normal through each pole. The meridians near the pole will go as the
meridians of this ellipsoid, as constants times $\sqrt{z}$. Such a function, damped quickly to zero beyond the singularities, may be sub-
tracted from the actual meridian curves at the poles, and the remainder can be fit by any conventional method of splining.

Then the full data by section, together with the splined data on
meridians, can be lofted by a Coons technique of bicubic patches (cf.
Forrest, 1972) to reconstruct a smooth surface in E^3 which goes by
explicit formula within coordinate patches. For an algorithm, see
Rogers and Adams, 1976: ch. 6. In particular, the curvature at any
point of the surface may be extracted by formula without any further
reference to the original data. If the original data are finely
enough measured, one will not need to spline at all; one can tri-
angulate in between successive slices using a method such as that of

Fuchs, Kedem, and Uselton (1977), then estimate the curvatures by the method of Gauss himself, by principal components of the scatter of the unit normals at the centers of nearby patches.

A simple extension of this device may be useful for interpolation between two surfaces with homologous landmarks located on each. Let each surface also be sliced, as above, the landmarks not necessarily falling upon any of the slices used. The slicings define the shapes of the surfaces separately, while the landmarks provide information about the homologies of regular points everywhere thereupon.

When the landmarks are projected onto the coordinate cylinders, upon which we lofted the surfaces via Coons' techniques, there results a map which can be unrolled onto a plane. We have two such maps, one of the coordinates of the landmarks of image 1, one for image 2. The one-to-one correspondence between these sets of landmarks may be extended, by the methods of Tobler (1977), to produce an interpolated one-to-one correspondence throughout the cylindrical projected planes. To interpolate from one smoothed surface, A, to another, B, one would proceed in three stages: first, project from A onto one of its coordinate cylinders, by numerically inverting the lofting; second, by planar interpolation among projected landmarks, find a corresponding position in a coordinate cylinder for surface B; third, loft from B's cylinder back up to the surface B by the formalism of the appropriate Coons patch.

C. Skeletons

Returning now to plane problems, I shall seek another sort of solution entirely. Application of the notion of curvature has reduced the measurement of shape to a series of observations upon boundary segments, their lengths and net turning. This can be depicted as a set of vectors chasing from landmark to landmark around the boundary.

Now consider the problem of measuring a wiggly worm, Fig. IV-11 (a). It is clear that the intrinsic approach is not particularly helpful here. What is visually crucial to the extraction of measures from this shape is the correspondence of the opposite sides. This is a sort of width perception relating boundary arcs all along the parallel sides. If the worm straightens out some, our intuition expects that the width of its form will stay practically the same, although the pairing of boundary points has in fact altered. We also expect its length to be invariant. Length here means arc-length down

the middle, along the arc $\big\backslash$ for the wiggly shape and $\diagup$ for the
less wiggly. As the worm curls up in a circle, the Euclidean distance between its ends becomes a smaller and smaller fraction of the
intuitive length of the figure, as in Fig. IV-11(b). We need a way
of measuring that somehow traces out the intuitive middle in this
sense and lets us estimate a "width" in curvilinear fashion. It
will thereby generate a correspondence of boundary points, in con-
tradistinction to the conventional approach, which must pair them
into endpoints of segments in advance.

Such a procedure was invented a few years ago by Blum, who has
reviewed its history in Blum (1973). Though it represents a basic
geometric intuition into problems of shape, it is nearly unknown outside the picture-processing profession. It is worth setting forth in
some detail, since expositions are scarce.

The skeleton of a plane figure is a certain graph inside the
figure, together with a function on the graph. There are several
equivalent definitions, of which the following is perhaps the most
concise: the skeleton is the locus of all points which do not have
a unique nearest boundary point upon the shape; the function is the
distance to any of the set of equally distant nearest boundary
points. Alternatively, one may characterize the skeleton by the
"grassfire" model. We imagine a shape boundary to be "drawn" on the
dry grass of a prairie, and fired. The fire will burn evenly in all
directions from its starting locus until it encounters points at
which it arrives simultaneously from two directions, whereupon it
quenches itself, as grass cannot burn twice. Such loci comprise the
skeleton, and the function we seek is the time it takes the fire to
arrive there and go out. The reader may test his understanding of
this concept by verifying the equivalence, too, of the next two
definitions: the skeleton is the set of points at which the surface
whose z-coordinate is distance-to-boundary fails to have a tangent
plane; the skeleton is the set of points on no other point's
shortest-path-to-boundary. Figures IV-11(c), (d), (e) show various
examples of skeletons; Blum (1973) and Blum and Nagel (1977) present
many more.

Nomenclature for the skeleton is not standardized. Some authors
call it the medial axis, others the symmetric axis. The skeletal
function is variously called the quench function, the medial axis
function, and the symmetric distance. The two together are often
called the skeletal pair. Part of the skeleton is a connected graph

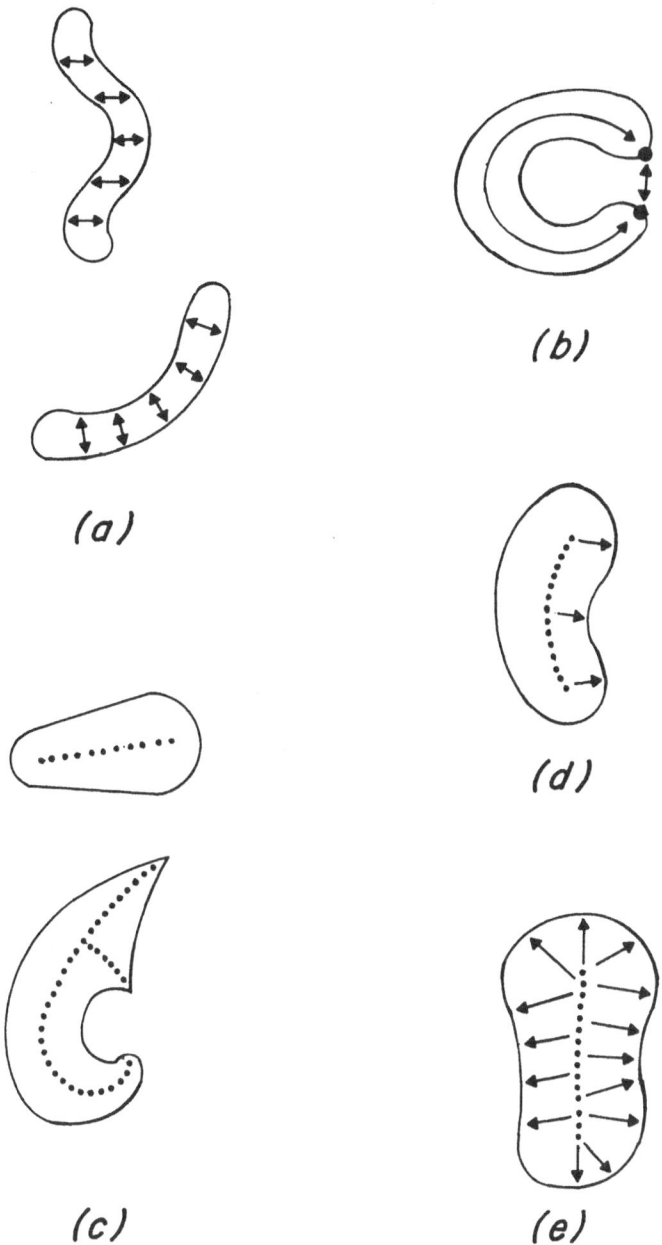

Fig. IV-11.--On skeletons. We need measures of (a) width, and (b) length, which do not change when a form bends. (c) The dots are on the skeleton or medial axis, the locus of points with no unique nearest neighbor upon the outline. (d) Skeleton of a kidney bean, with a suggested parameterization. (e) The complex-valued functional exhaustively representing the skeleton, and thereby the outline: directed distance to boundary.

inside the boundary. In addition, whenever the original shape is not convex, there will be disconnected branches of the graph beginning some distance outside the boundary and heading out to infinity. We ignore those branches, as they bear no new information.

From the skeletal pair one can fully reconstruct the original point set. About each point of the skeleton one places a disk of radius equal to the skeletal function there; then the total plane region covered by these disks, regardless of overlap, is exactly the original shape. Therefore the skeletal pair is an exhaustive representation of the shape, just as the intrinsic curvature function is, and we may with impunity attempt to extract features from it.

Blum has proved that skeletons flex the way the worm does. When we change the curvature of the skeletal arc without altering the quench function, there results a related shape with exactly the same perimeter and area as the original. In particular, the skeleton may be straightened out entirely. There results a symmetric shape canonically related to the earlier one--it is its width profile. The curvature of the skeleton is the average of the curvatures of the boundary arcs it matches. Even with data of wider variation than human crania, which were the implicit subject matter of the tangent angle method, the skeleton will correspond fairly closely to a "main axis" if there is one. It is an ideal descriptor for data of few landmarks. The curvature trace of a kidney-shaped object, for instance, is difficult to use as a template. It has no clear extrema or landmarks. But the skeleton of a kidney-shaped object, sketched in Fig. IV-11(d), is much more characteristic in form. It has two ends and a middle, all well-defined, and a skeletal function and skeletal arc both in the form of a cup. Such a set of skeletons is easy to parameterize.

A multivariate statistical method. Let us assume that the population of shapes under study is sufficiently regular that the topology of the skeletons is invariant. Each branch of this common skeletal graph is a simple curve which is fully defined by two functions along it: the skeleton function and its own curvature. We can now complete the picture of a "moving width" imagined above. From the two scalar fields on the skeleton we can construct the vectors skeleton-to-curve as they pass from end to end, as in (e). This provides a serial representation of the original shape in between the coordinate representation (Cartesian displacements from a fixed center) and the intrinsic representation (elements of arc all the way around the boundary).

It would be good to pass this vector-valued function, sampled at intervals along the length of the skeletal arc, through a multivariate analysis. As with the treatment of the closure criterion ʻin analysis of curvature, a modification of standard statistical methods is needed to come to terms with the particulars of the application. In this case a reasonable adjustment, it seems to me, is the use of complex covariance matrices. We sample the displacement function at intervals calibrated by skeletal arc length or skeletal landmarks, and consider each displacement a complex number instead of a pair of real variables. The two scalar functions characterizing the skeletal pair are implicit in this single complex-valued function. The variance of such a variable z over populations of shapes is now the form $N^{-1} \sum zz* - \bar{z}\bar{z}*$, where N is the sample size and $*$ is the operation of complex conjugation. Its value is the sum of the variances of the x- and y-components separately, and is invariant under Euclidean congruent transformations. Covariances are now complex numbers, embodying a rotation as well as a magnitude.

Given a list z_1, ..., z_n of random variables in the complex plane such as these displacements, one may think of the vector space of linear combinations of these variables, forms $\sum w_i z_i$ for complex constants w_i. The notion of principal components generalizes to complex variables as follows: given a set of variables z_i of covariance matrix $S_{ij} = N^{-1} \sum z_i z_j* - \bar{z}_i \bar{z}_j*$, to find the vector $(w_1, ..., w_n)$ for which $\sum w_i w_i* = 1$ and var $(\sum z_i w_i)$ is maximal, one extracts the first principal component of the matrix S. The component has positive latent root, since S is Hermitian positive-definite. Cf. Rao (1973:42). For the actual extraction of such eigenvectors, see Wilkinson and Reinsch (1971:197).

Successive principal components, corresponding to uncorrelated scores of successively maximal variance, can be extracted and used in factor analysis, discriminant analysis, canonical analysis, and multivariate analysis of variance. There is then no need to standardize the orientation of the objects measured: they will be classified by the amplitudes $(\sum w_i z_i)(\sum w_i z_i)*$ of their complex scores. In the same way that from a factoring of $\quad\overline{}\sim\quad$ -type curves there emerge pure "profile" factors, from a factoring of the covariance matrix of vector displacements will emerge pure shape factors, systematic bulges and flexes in the diverse parts of the shape. These factors will be clearer if we change bases to involve the differences $z_i' = z_i - z_{i-1}$ of successive vector displacements, or the complex ratios $z_i'' = z_i/z_{i-1}$, either of which serves as surrogate for differentiation of the original series. The transform z_i''' $= \log (z_i/z_{i-1})$ corrects for both orientation and scale. In this form of analysis, a section systematically curving like $\rightsquigarrow$ will be represented in a factor with equivalent weight whatever its orientation to the rest of the skeleton. For instance, a worm of constant width turning with constant curvature will have all its displacement vectors on a circle and all its ratios constant: $\overset{\uparrow\uparrow\uparrow\uparrow}{\frown}$ yields $\text{\textbackslash}\psi$. A bulge gives a pattern of reflection of difference scores at the extremum: $\overset{\uparrow}{\frown}$ yields $\downarrow$ or $\wedge$. Dents look like $\text{\textbackslash}\psi$, and so on. If the z_i constitute a fair and evenly spaced sampling of the skeleton between its landmarks, nodes, ends, and maxima of curvature, there results a series of ideal shapes, corresponding to the explicit loadings on each factor, which can be drawn, superimposed, and contemplated. Different branches of the skeleton correspond to different lists of z's which can be analyzed separately or together in large analyses. One can compare thereby the systematic shape variations of separate lobes and of the whole.

Bibliographic note. The literature exploring the implications of Blum's seminal idea is unaccountably lean. The picture-processing people encountered his work via symposiums and recognized the elegance of the idea; in a flurry of articles in the late 1960s they published algorithms for its computation from digital data. Cf. Rosenfeld and Pfaltz (1966) and Montanari (1968, 1969). To my knowledge, only Oxnard (1973) has published a scientific inference established by the study of skeletons. The skeleton has been generalized to grey-level pictures, in which the "velocity" of the grass-fire is inversely proportional to picture density, by Levi and Montanari (1970), and, somewhat differently, by Hilditch (1969), wherein it is applied to the analysis of chromosomes. Recently Moore (1974) has espied a connection between skeletons and integral geometry. The challenge of a suitable metric for skeletons, posed by Calabi and Hartnett (1968), has not been addressed since.

Recently the ALGOL source code has been published for a fairly simple algorithm which produces skeletons as unordered point-sets, but it does not examine their connectivity or continuous geometry in any way (de Souza and Houghton, 1977). Blum and Nagel (1977) sketch a quantification of the skeletal pair (quite different from that presented here) emphasizing discrete properties (signs of curvatures, connectivity at nodes) weighted by the expanses of boundary associated with the arcs separately. The generalization of the skeletal approach to three dimensions lies fallow, but there is a tantalizing hint in Agin and Binford (1976) for structures with sensible limbs.

By way of closing this exploration into geometric methods of shape measurement I wish to reassert the methodology I propose. The existing lore on statistical shape measurement hews to the principles of Euclidean index construction: distances, angles, areas, ratios, all among preassigned landmarks. I have suggested that the shape be treated instead at all times as an extended locus in real Euclidean space of two or three dimensions. Immediately certain powerful geometric analyses come into play, and we have available certain continuous functions (serial representations) which wholly encapsulate the original shape in a coordinate-free numerical form. The intrinsic equation of a curve in terms of its curvature is one such representation; the skeletal pair is another such, remarkably different in appearance but exactly as informative. By sampling from these functions and analyzing populations of shapes using appropriately modified statistics, we avoid the severe methodological constraints of conventional landmark-based multivariate morphometrics, the extraction of finite distances and the passage to merely linear statistics before any aspect of curved form has been allowed to express itself. I argue for a geometric morphometrics in which multivariate statistics enters solely to compress the information extracted from a delicate analysis of actual curved form.

SECOND PART. THE MEASUREMENT OF SHAPE CHANGE
USING BIORTHOGONAL GRIDS

CHAPTER FIVE. THE STUDY OF SHAPE TRANSFORMATION
AFTER D'ARCY THOMPSON

The boundary represents the shape, but it is not the boundary
which grows. A rubber band released from tension happens to return
to its elongate resting shape. In the course of release, the band
is active, having its own preferred curvatures. There may exist
biological situations in which an analogous determination occurs,
but in the vast majority of shape changes the boundary is passive.
It is pushed out or pulled in, pressured hither and thither by the
changes everywhere inside of the shape. The inside, in growing,
carries the boundary along with it, embedded in whatever distortions
are wreaked about it. Intuitively we know what the distortions have
to be like: smooth correspondences with no tears or folds, no sud-
denly disappearing or appearing tissues.

Two methodological problems arise in connection with the use of
distortion as a model for biological phenomena. First, material may
not be conserved in growth processes even where form is conserved.
Craniofacial bones grow by remodelling, for instance, and implants
eventually fall out during extended development. The implants, however, are not the landmarks whose homology we seek to extend.
Rather, the constancies of shape, as controlled by the periosteum or
a functional matrix, generate a correspondence between corners, foramina, and the like regardless of individual cell histories. Other
structures may grow by accretion, likewise "creating" landmarks and
destroying the role of others. In this case, too, our attention
moves with the form, not the substance.

Second, biological tissues do more than change their shape.
They rearrange themselves discontinuously, they die or are replaced,
they differentiate. The study of shape, as I have defined it, ignores all these processes. It does not presume that they are absent,
only that they are corrected for, by an analytic focus upon geometrically comparable structures. It is sometimes not clear whether a
certain comparison is a shape change or not, that is, whether the
sole relevant discrimination between corresponding parts of two organisms is spatial position. In particular, all morphometric methods
must beg the problem of homology versus analogy. Even in ontogenetic

series, there is no necessary relationship between homology over
continuous time and geometric comparison of separated "snapshots."
The comparison of two forms may proceed by consideration of the
spatial arrangement of corresponding points, I submit, as long as
it provides useful information for subsequent explanations and
theories. Technical failures of homology, such as are described
in the preceding paragraph, do not necessarily require abandonment
of the geometric level of analysis in ontogenetic series, phylogenetic series, or populations.

Now in the analysis of shapes as equivalence classes of outlines I did not speak of internal points at all. Landmarks are by
definition on the outline, upon the outside edge. We know, though,
that in biological applications, when the outlines correspond so do
the interiors in some fashion, point for point. To describe shape
change in general, one must have--or must estimate--full details of
this interior correspondence. Without landmarks, we could say

nothing more about the relation between ⌂ and ⌂ than that the

first form has three lobes, the second four. Though the landmarks
express the correspondence of points, they do not describe the change
explicitly. They only summarize interior aspects of the whole change
in a way which can allow a great deal of ambiguity in extreme cases.

Consider the two different shape changes sketched in Fig. V-1
(a), (b). In one case, the two landmarks appear to have maintained
relationship with the whole, but an interior point "moved," dragging
the inside with it. In the other, landmarks kept their distance, but
the shape grew disproportionately on the sides of their join. Now
consider that these are the same shape change. I have simply rotated one diagram by 90° and chosen a different diameter for the
landmark pair. The complete description of the process, much more
adequately shown in (c), is via a Moebius (linear fractional) transformation upon the complex coordinate of the original figure (cf.
Schwerdtfeger, 1962: ch. ii). Let the radius of the left-hand circle
be 1, of the right-hand circle, k; let the center move to a point
k $(s/ \sqrt{(1+s^2)})$ inside the image circle, the point -1 move to -k, the
point +1 to +k. One complex transformation which takes these three
points into their three pre-assigned images is the map $z \rightarrow f(z) =$
$k(tz+s)/(sz+t)$, where $t = \sqrt{(1+s^2)})$. (See Pedoe, 1970: sec. 54.3.)
The "growth" described by this equation is an expansion evenly in
every direction at a rate equal to the magnitude of the derivative
of this function, which is $k/((sx+t)^2 + (sy)^2)$ for x,y the ordinary

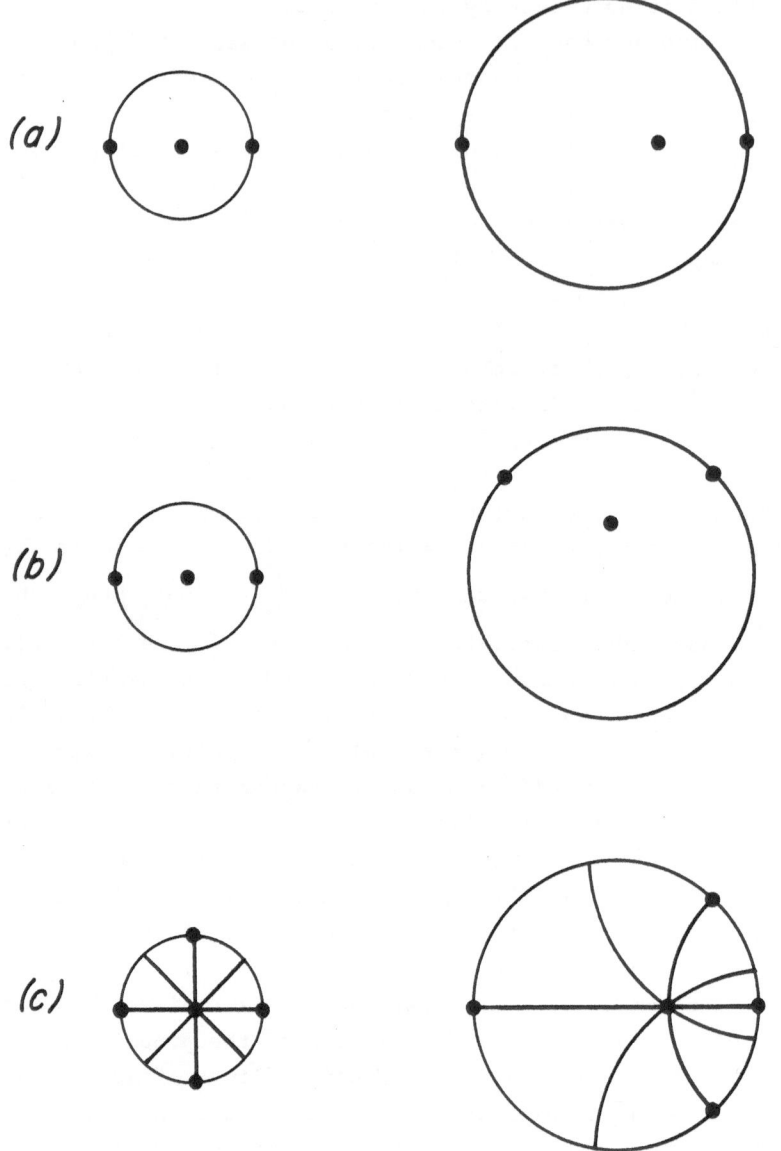

Fig. V-1.--(a, b) Two presentations (via different choices of landmark) of the same shape change, a Moebius transformation $z \rightarrow k(sz+t)/(tz+s)$ for suitable s, t, k with $s^2 = t^2 - 1$. (c) A clearer representation using both sets of landmarks and some additional diameters.

Cartesian coordinates of the smaller image. The crucial parameter
for this model is the movement of the center of the circle, and
that is not a landmark at all. (Its movement is hinted at in the
differential expansion of interlandmark intervals about the boundary.) Clearly, no pre-selected set of landmarks is necessarily
helpful in the description of change; we need to know where everything is going, to represent somehow an association between all
pairs of corresponding points in the two forms.

To represent the relationship between a pair of homologous
shapes throughout their interiors, D'Arcy W. Thompson in 1917 invented the Method of Transformation Grids. Thompson's brilliant
and elegant suggestion is that we construe the pair of diagrams as
a transformation of the whole picture plane which maps the points
of one diagram into corresponding points in the other while varying
smoothly in between. Half a century later his elegant inked examples from all the animal kingdom still evoke that happy intellectual astonishment which so often heralds long theoretical strides
in the sciences--but the reorientation of morphometrics has not come
to pass. All subsequent work still begins exactly where Thompson
left off. Yet, refractory though be its applications, his method is
a scientific advance of the first order. I will show that his work
is not mathematically perfected, that we can place more demands on
the method so as to arrive at a unique geometrical object. The
method remains, nevertheless, the first and only coming-to-grips
with shape change on its own terms. All subsequent serious work in
morphometrics represents analysis of the information-processing
problems engendered by the Thompson definition. All the information
we need is there, in one picture; we have but to learn how to report
it out.

A. The Original Method
1. Thompson's own work

The formal theme of D'Arcy Thompson's method is this: to represent a change of one shape into another by the single mathematical
object which is the map of one shape onto the other, and then to
visualize this mathematical object.

The preceding chapters of his essay had explored the applications of mathematical insight and geometrical models to various empirical forms and manifestations; the subject of shape change, to
which Thompson now turns, is just another abstractable aspect of

form. The virtue of mathematics is, after all, "to eliminate and
to discard; to keep the type in mind and leave the single case,
with all its accidents, alone . . . . [The] deformation of a complicated figure may be a phenomenon easy of comprehension, though
the figure itself have to be left unanalyzed and undefined. This
process of comparison, recognizing in one form a definite permutation or deformation of another, apart altogether from a precise and
adequate understanding of the original 'type' or standard of comparison, lies within the immediate province of mathematics . . . ."
(Thompson, 1961:271). His intention here, strictly methodological,
has two thrusts. The first is to continue his search for the basis
of form in force. "[The transformed representation] once demonstrated, it will be a comparatively easy task (in all probability)
to postulate the direction and magnitude of the force capable of
effecting the required transformation" (ibid.:272). This goal is
now considered quite archaic, as causation is now universally considered to work through natural selection at the level of ontogeny
(Gould, 1971:271f.). But the other thrust is still cogent. Thompson sets two conditions: "that the form of the entire structure under investigation should be found to vary in a more or less uniform
manner, after the fashion of an approximately homogeneous and isotropic body . . . , and that our structure vary in its entirety, or
at least that 'independent variants' should be relatively few"
(Thompson, 1961:274). These postulates accord with his belief that
correlation of characters is the rule, that constituent parts of an
organism never can evolve quite independently. From this he draws
a sweeping conclusion:

> When the morphologist compares one animal with another,
> point by point or character by character, these are too of-
> ten the mere outcome of artificial dissection and analysis.
> Rather is the living body one integral and indivisible.
> whole, in which we cannot find, when we come to look for
> it, any strict dividing line even between the head and the
> body, the muscle and the tendon, the sinew and the bone.
> Characters which we have differentiated insist on integrat-
> ing themselves again; and aspects of the organism are seen
> to be conjoined which only our mental analysis had put
> asunder. The co-ordinate diagram throws into relief the
> integral solidarity of the organism, and enables us to see
> how simple a certain kind of correlation is which had been
> apt to seem a subtle and a complex thing.
> But if, on the other hand, diverse and dissimilar
> fishes can be referred as a whole to identical functions
> of very different co-ordinate systems, this fact will of
> itself constitute a proof that variation has proceeded on
> definite and orderly lines, that a comprehensive 'law of
> growth' has pervaded the whole structure in its integrity,

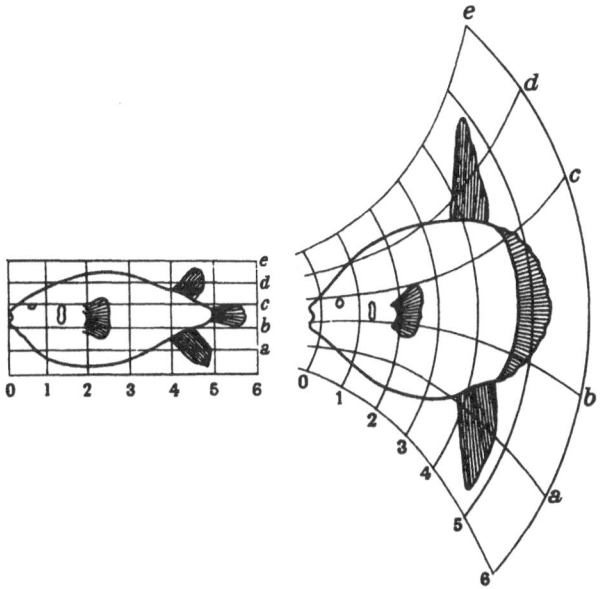

Fig. V-2.--Cartesian transformation from <u>Diodon</u> to <u>Orthagoriscus</u>
(= <u>Mola</u>). From Thompson (1961:301).

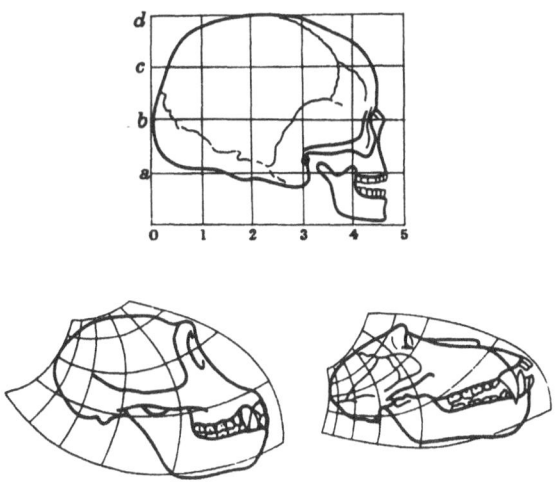

Fig. V-3.--Cartesian transformation from "human skull" to chimpanzee and to baboon. From Thompson (1961:318-9).

and that some more or less simple and recognisable system
of forces has been in control. It will not only show how
real and deep-seated is the phenomenon of 'correlation',
in regard to form, but it will also demonstrate the fact
that a correlation which had seemed too complex for analysis or comprehension is, in many cases, capable of very
simple graphical expression. (ibid.:275-6)

Abandoning, then, his search for physical causation, in view of

the complexity of the situation, Thompson sets out on a "floating

mathematics for morphology, unanchored for the time being to physi-

cal science, but capable of valid generalisation on its own level"

(Hutchinson, 1949:579). He contents himself with a large variety of

examples drawn from all over the two living kingdoms. Of these, the

analysis linking Diodon and Orthagoriscus (= Mola) is probably the

most famous of all. His analysis is quite instructive. (I have

changed his figure numbers to correspond to mine.)

Figure V-2 (left) is a common, typical Diodon or porcu-
pine-fish, and in Fig. V-2 (right) I have deformed its vertical co-ordinates into a system of concentric circles, and its
horizontal co-ordinates into a system of curves which, approximately and provisionally, are made to resemble a system
of hyperbolas. The old outline, transferred in its integrity
to the new network, appears as a manifest representation of
the closely allied, but very different looking, sunfish,
Orthagoriscus mola. This is a particularly instructive case
of deformation or transformation. It is true that, in a
mathematical sense, it is not a perfectly satisfactory or
perfectly regular deformation, for the system is no longer
isogonal; but, nevertheless, it is symmetrical to the eye,
and obviously approaches to an isogonal system under certain
conditions of friction or restraint. And as such it accounts,
by one single integral transformation, for all the apparently
separate and distinct external differences between the two
fishes. It leaves the parts near to the origin of the system, the whole region of the head, the opercular orifice and
the pectoral fin, practically unchanged in form, size, and
position; and it shows a greater and greater apparent modification of size and form as we pass from the origin towards
the periphery of the system.
In a word, it is sufficient to account for the new and
striking contour in all its essential details, of rounded
body, exaggerated dorsal and ventral fins, and truncated
tail. (Thompson, 1961:300-1)

A close reading of his own diagrams unearths several inconsis-

tencies. The throat of Mola is the mirror-image of the area dorsal

to the eye, but not so for Diodon; yet the grid is symmetrical, and

hence has not accounted for this clear dorsoventral difference.

The divergence of the outermost horizontal grid lines of Mola begin-

ning at the far left, why does the divergence of the axes b, c begin

only behind the pectoral fin? The caudal margin of the anal fin has

shifted between diagrams from axis b to halfway between axis b and axis a, so that the vertical expansion of the axes rightward in Mola is insufficient to match the data. Should we reverse the roles of the two genera, so that Mola supported the rectilinear grid, what then would be the reading of the transformation? Would it still be made of conic sections, circles and hyperbolas? Thompson left no notes on the detailed construction of these drawings, so we cannot tell how he settled upon the axes that actually appear.

A subsidiary theme emerges in the course of Thompson's survey which has proved of most enduring interest among later students of the method. This is the diagrammatic sequencing of transforms and its cognate in paleontology, the search for intermediates. The most successful example is the very regular succession of deformations in the phylogeny of Equus. Thompson's interpolation is by drawing fractional parts of grid intersection displacements between figures. It is very simple, and takes no account of possible regional varia-tion in evolutionary rate, but in this case it works beautifully. More famous than this is the less successful demonstration that the distortion from human to baboon differs "only in an increased in-tensity or degree of deformation" from the distortion of human into chimpanzee. His figures are assembled in my Fig. V-3. "In both di-mensions, as we pass from above downwards and from behind forwards, the corresponding areas of the network are seen to increase in a graduate and approximate logarithmic order in the lower as compared with the higher type of skull; and, in short, it becomes at once manifest that the modifications of jaws, brain-case, and the regions between are all portions of one continuous and integral process" (ibid.:319-20). It is unfortunate that Thompson inexplicably began with a "human skull" of a braincase impossibly large; in addition there are errors of drawing similar to those for the Diodon example, for instance the wandering of the gonial angle relative to the lower end of coordinate line 4. Thompson hoped to fill in this series too with intermediates, but the necessary fossil material was unearthed only after he ceased work. Without saying quite what he means by a "direct line of deformation," he concedes that neither of the apes lies "precisely" in the sequence of the other's hypothetical con-nection with man.

It seems that Thompson's postulate of homogeneity took prece-dence over his draftsmanship in most of these examples, and it is not surprising that grids published after him, with the exception of one of Needham's, Fig. V-6, are invariably less legible.

2. Later examples

It is fair to say that, after Thompson's original publication, the method we are discussing underwent "vicissitudes" rather than "development." It has mainly been the last theme, inter-relating the grid method with techniques of ordination, that accounts for most of the true grids computed by persons other than Thompson. A good review of this literature to the early 1950s may be found in Richards (1955). No new examples at all appeared until the 1930s, perhaps owing to Thompson's failure to provide instructions. Colbert (1935) displayed a "Cartesian coordinate chart to illustrate the manner in which the skull of Rhinoceros might have evolved through Gaindatherium from a primitive form such as Caenopus . . . . This is essentially the method used so widely by D'Arcy Thompson, but here it is applied in a more detailed manner than was done by that author." The drawings are excellent, but the accompanying text runs only ten lines. Colbert seems to have come to no useful conclusions at all for his pains.

A major difficulty for Thompson in his drawings was the location of grid lines traversing large regions without landmarks, such as the cranial vaults in Fig. V-3. A clever solution to this problem was put forward by Avery in 1933. He inked a square grid on a small tobacco leaf and photographed it over its subsequent development. There resulted an empirical sequence of true Thompson transformations, from which he extracted directly the areas of the little grid-quadrilaterals and turned the whole into an analysis of areal growth-gradients after the fashion of Huxley.

Much more thoroughgoing in its use of geometry is a subsequent reanalysis of Avery's data by Richards and Kavanagh (1943, 1945). The leaf images, the points followed individually throughout the growth, are exact enough to support a differentiation everywhere. From the derivative one may ascertain just what growth there is in any direction, and the authors present the directions of maximum and minimum growth by little crossed axes, of length proportional to directional growth-rate, scattered over the leaf figures. It is not clear to me how these were computed, as the derivatives postulated by their model need to be estimated with care whenever the current system is not Cartesian, which is to say, at all times after the instant of the tattooing. Tobler (1978) has recomputed these axes, still using Avery's original data, and shows unsystematic errors in Richards and Kavanagh's computations scattered throughout the images.

Criticism of the arithmetic should not obscure the advance those authors were explicitly attempting toward a tractable transformation theory. They consider their method a combination of the Cartesian transform with the method of growth-gradients to provide both numerical and geometric insight into observed process. Their conclusion, in particular, is quite Thompsonian in spirit:

> It seems reasonable to expect that the pattern of specific growth-rates should be explicable in terms of the characteristics of the growing material at the various points in the organism . . . . In the example of the tobacco leaf, there was a tendency for the direction of maximum rate to coincide with the direction of the vascular bundles. It may be worthwhile, in future investigations, to see how closely the directions of these maximum and minimum rates can be related to recognizable structural characteristics of the organisms.
>
> In cases in which the directions of the extreme rates prove to be tied in this manner to definite structures within the organism, the problem of describing the growth transformation becomes somewhat more definite. At any instant the directions of extreme rates determine throughout the organism three families of surfaces such that any member of one family intersects any member of the other families at right angles. (The simplest example of such a system of surfaces is, of course, the set of planes determining a rectangular coordinate system. In the case of growth in a plane, the directions of extreme growth-rate determine two mutually orthogonal sets of curves.) As the growth continues, these surfaces, though transformed, will still remain at right angles to each other. (Richards and Kavanagh, 1945:228-9)

There never were any "future investigations," and these facts had to be rediscovered by me in the course of my own researches (see chapter vi). The geometry of this analysis has been imitated only once, and then only with the abandonment of the original impetus, the dependent variable having been converted from a shape change to a field of scalar growth rates. Erickson (1966), demonstrating a computerized version of the calculations, neither makes any use of the directions of growth (only of its directionality, a scalar) nor cites Thompson in his references.

Other studies, unable to deal with descriptions of single grids in extenso, have explored the ordination parameter in Thompson's seriation scheme. Kummer (1952) attempted a consistent reconstruction of the hominid line from Proconsul through modern man, then fit many intermediate fossils to fractional positions along this transformation. The brilliance of the drawing is tempered somewhat by the goodness-of-fit to Piltdown Man and by his evident disdain for the possibility of regionally varying rates of hominization. He nowhere explains how he estimates the serial position of particular forms, nor why the distances all happen to be integral multiples of

one-eighth. Lull and Gray (1949), also using Thompson's device for
linearized interpolation, find the coordinate method too sophisticated for the available data on ceratopsians (a family of dinosaurs
including Triceratops). Their extrapolations buckle the plane,
growth-gradients become apparent too soon, and the transformations
of ontogeny are not related to those of phylogeny in any compre-
hensible way. They conclude that nothing new about ceratopsian
phylogeny can be discovered by this method, which ought not to replace the traditional system of "comparing single pairs of measurements." Yet they conclude, oddly, that the fault is in the data,
not in the method: the Thompsonian grid is too attractive to be re-
jected.

The technique has also been used to convey negative findings,
to espy differences rather than similarities. Starck and Kummer
(1962) drew out human and chimpanzee cranial ontogenies in parallel columns of a figure, then argued that their differences were
too complicated for summary in the famous "fetalization" hypothesis
that Homo is but a chimp whose development is retarded, whose fetal
growth rates reign much longer into infancy. I will re-analyze
this matter in chapter vii. In Kummer (1953) can be found analogous
series for other primates and adumbrations of their similarities and
differences. (The detailed construction of the drawings has been
criticized severely by Helmuth, 1970.) For reptiles, Olson (1975)
displays sixteen grids showing distortion of various fossils in a
grade away from one of their number. He concludes that although the
conventional ordination of these forms is misleading, trends buried
under specialized adaptations, one can select three forms which
"lie close to the phyletic line" and then construct intermediates,
the better to characterize the specializations. These conclusions
are not founded upon any algorithms for objective diagnosis of the
grids displayed.

Several studies have attempted to resolve the perplexities of
the grid method by the application of some other statistical technique. In 1967, Sneath published a method which, though called
"trend surface analysis of transformation grids," is in fact a somewhat different product for the analysis of displacement trend-surface grids. Sneath's interest is not in the geometric features of
the grids themselves, but only in the "factors underlying the deformation" and particularly in the gross differences in general shape.
His purpose is taxonomic, not biotheoretic; he wishes taxonomists
not to be fooled by repetitious correlated expansions that could be

characterized by a few coefficients of a function. He borrows from geology the notion of a trend-surface, which summarizes scalar data distributed over a map. Operating from an arbitrary sample of corresponding points in pairs of images optimally superimposed, he fits separate trends for the two Cartesian displacements, vertical and horizontal, of the points in one diagram relative to their mates in another. He then simply partitions the trends into polynomial components and compares coefficients from fit to fit, for his intent is not, after all, the mapping of shape change. In spite of its basis in spatial data, this technique has no specific geometric or biological content. Components of the trend (i.e., quadratic, cubic, ..., terms) link together discrepant curves and bends throughout the observed image, in disregard of biological structure. All local phenomena are smoothed out; in particular, there is no way of estimating growth-rates in the various parts of the image. We are unable to return from the statistical manipulations to the original direct transforms of shape.

Oxnard (1973:62-6) curiously reverses this tactic. He uses transformation grids to summarize some canonical axes produced by conventional multivariate analysis. Having extracted three main factors discriminating primates by their scapulae, he selects three pairs of scapulae which differ only on a single score, first, second, or third. He draws the resulting transformations and looks for stresses which correspond to variables heavily loading on the factors. One grid is a "cranio-lateral twist," as in Fig. V-4; another is a "mediolateral compression," and the third a "craniolateral stretching." Since factor scores are interval measures, it would be very fine if these categories of deformation were also. Can we ask how much cranio-lateral twist is in the transformation shown, and can we verify that reversing the direction of the transformation gives us the inverse amount of depicted twist? It would be very exciting if the transforms added or multiplied somehow to account for comparison of scapulae differing on all the scores. But, as published, the words correspond to no formal properties and do not suggest any possibilities for measurement.

3. Difficulties

Thus the vicissitudes of the explicit method of transformation grids since its inception. Those who have attempted to make of the technique a tool for analysis of understanding commensurate with the

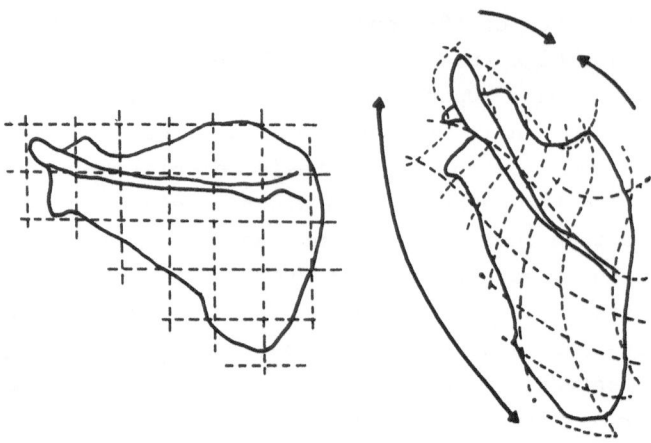

Fig. V-4.--Cartesian transformation from scapula of Papio to scapula of Gorilla, with indication of "cranio-lateral twist." After Oxnard (1973).

elegance of its illustrations have often superimposed some arithmetic field over the original design for the engridment. Under this rubric I would include the elemental growth-rates of those who studied Avery's leaf, and also the Cartesian displacements of Sneath. The others, those who did not add quantity to Thompson's original diagrammatic scheme, generally have failed to gain any geometric insight either, however carefully drawn their diagrams. I include Thompson himself in this latter category. In all his analyses, he rests with the possibility of a single system of forces--not a very well-defined analytic category. His goal was not to measure; he was content to exemplify the geometry logically prior to any measure.

The problem here is fundamental. It seems impossible to extract quantity from the Cartesian grid, as Thompson formulated it, in any straightforward way. Even after a decade during which the brightest graduate students all have had access to computing power adequate for large multivariate data sets, there is no hint in the literature of a line of attack upon quantification once one has painstakingly drawn out the Cartesian grid. Sokal and Sneath (1963:82-83) argue from two quite well-drawn examples of accurate, intentionally incomprehensible diagrams that the problem is one of feature enumeration. For any "realistic" grid fitting the data more closely than Thompson's (which is not a difficult accomplishment), various ebbs and

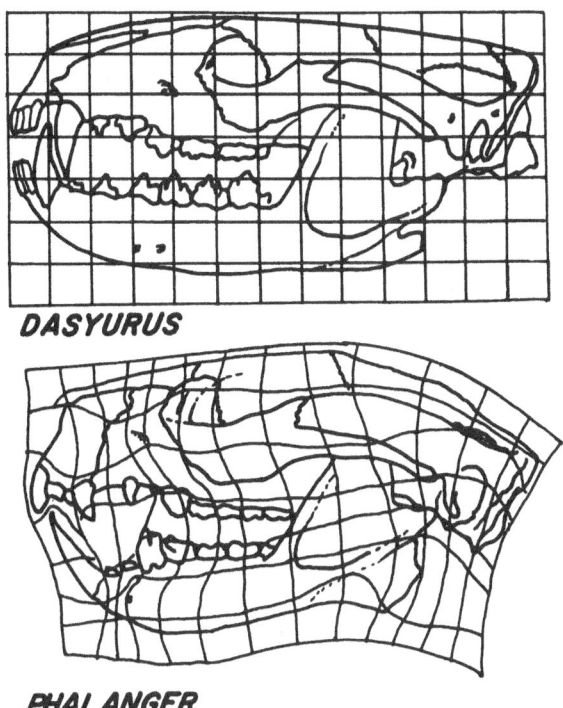

DASYURUS

PHALANGER

Fig. V-5.--A sample realistic Cartesian transformation. After
Sokal and Sneath (1963).

flows of the lines become apparent; it is clear that more than one
source of curvilinearity is at work, and these represent, in turn,
a multiplicity of sources of variation. We have no visual facility
to count these elementary "fields" or sense how they could be separated and measured. In the effort to talk about what is there we
open our mouths and become speechless. There are stretches in certain slowly varying directions, and certain subtle changes of angle
overall, and a constant shifting in the relative spacing of grid
lines in both directions. The visual complexity of these grids is
frustrating and indescribable, like distortions in an unflat mirror
whose shape we cannot comprehend. Figure V-5 exemplifies all these
difficulties.

This is what Medawar meant when he called the method of transformations "analytically unwieldy" (Medawar, 1958:231): that the
transformations cannot be analyzed, broken into parts. It is very
well to declare that a single shape change is all of a piece, but if

families of them have to be analyzed, then we need some means of deciding when a whole collection is likewise all of a piece. To do
this we need to count the pieces. When the data are as complex as
Lull and Gray's ceratopsians, the methods which succeed elegantly
for the equids fail utterly. Where are we to turn?

The compromises I will review in the next sections all determine in advance of the analysis certain distances or separations
from whose statistical manipulation a grasp of structure is expected to emerge. It seems to me that no such method can be general, can be validly applied to more than a handful of data sets.
I will argue cases as I explain them, only to conclude that we have
to start over.

B. <u>Analysis of Growth Gradients</u>

In a great many samples of related shapes, pairs of distance-
separations (and other types of variables as well) obey an equation
of the form $y = bx^k$, the <u>allometric</u> equation, which is to say that
they fall on a straight line in log-log plots. Such behavior will
obtain whenever the intrinsic growth rates $(dx/dt)/x$, $(dy/dt)/y$ of
the two separations are in constant ratio. In the special case in
which one of the variates is general body size (measured by overall length), we are studying the change of proportions with size.
Gould (1966) would have this be the definition of allometry, irrespective of constancy of coefficients; but it is better for a discussion of Thompson to accept Huxley's original definition (elaborated
in Huxley, 1932), in terms of numerically constant or slowly changing differential growth ratios among diverse parts and dimensions.

When a large number of short segments throughout a collection
of organisms show constancy of pairwise relative growth rates during development, Huxley would have us speak of "the general distribution of growth-potential," or <u>growth-gradients</u>, which are the
various coefficients k, growth rates with respect to some standard
rate, distributed spatially over the organism. When these ratios
are not quite constant, but vary smoothly with position and biological age, we may still compute the gradient field as it varies over
successive growth periods.

In his original publication, Huxley interpreted his method as
a quantitative refinement of Thompson's original scheme. Referring
to the <u>Diodon</u> transform, Fig. V-2, he writes:

> If, as D'Arcy Thompson points out, the transformation, so
> difficult to understand at first sight, becomes readily compre-

> hensible on the idea of an orderly change in the distribu-
> tion of growth-activity along the axis of the body, then
> clearly the proportions of the animal must be continually
> changing so long as it is increasing in absolute size, or
> at least over a long space of time. But the fish's outline
> and the system of co-ordinates drawn to fit it represent
> the state of affairs only at one particular moment of its
> life-history. If the fish had grown to twice the bulk, its
> proportions would have changed, and the co-ordinate grid
> would have to be altered; yet the underlying growth-grad-
> ient might have remained wholly unaltered . . . .
>   For this reason, the co-ordinate method, while of the
> utmost importance as affording a graphic and immediate proof
> of the need for postulating regularities in the distribution
> of growth throughout the body, is of little use for detailed
> analysis, because by its nature it neglects the fundamental
> attribute of differential growth, namely the change of rela-
> tive proportions with absolute size: it is static instead
> of dynamic, and substitutes the short cut of a geometrical
> solution for the more complex realities actually underlying
> biological transformation. (Huxley, 1932:105-6)

Apparently Huxley never noticed what emerges from four decades'
hindsight: the method of gradients requires that the directions in
which the gradient is likely to go be specified in advance--a "geometrical" input. He knew that if separate parts of a limb manifest
a nontrivial growth-gradient, then the whole length is necessarily in
exact allometric relation with none of its parts. He did not notice
that if growth proceeds allometrically with different coefficients
in two different directions from the same point, then even though
total area is growing allometrically, displacements in arbitrary
directions are not. A rectangle growing as x^m along one side and as
x^n along the other will have a diagonal growing as $\sqrt{(x^{2m}+x^{2n})}$, which
is not in allometric relationship with either side. I have already
discussed the modification of this system by Richards and Kavanagh
to account for directionality; the result is still expressed in
terms of scalar fields--rate of increase in area, anisotropy--not
leading to any major advance in the praxis of data analysis.

     In practice, then, the method of growth-gradients has been ap-
plied only when the axes along which gradients are to be measured
are constant and fixed in advance. In the commonest instance, there
is only one dimension of extent to be had, so that the gradient is
necessarily a function of one real variable, a particularly simple
sort. A locus classicus of this style is the work on segmental
lengths of arthropod limbs (Huxley, 1932; Needham, 1937, 1943;
Teissier, 1960). Medawar (1945) simplified the algebraic treatment
of these unidimensional gradients by introduction of an explicit
function for the relative position of any landmark as a function of

age. (Yates, 1950, showed by a reanalysis that Medawar's data, human vertical proportions, do not support the verbal interpretation he placed upon them; but this is not the fault of the formal innovation.) For systems without persevering landmarks, such as growing plant tissues, the necessary mathematics is more complex: age of the specific tissue must be entered as an additional variable. Salamon, List, and Grenetz (1973) exemplify this more sophisticated analysis for the "streak photograph," a data-collection device which represents differentials explicitly by divergence of neighboring streaks, and Green (1976) provides simple equations which further take the cell cycle and cell partitioning into account.

In systems of more than one intrinsic spatial dimension, there are two methodological possibilities. In the study of accretionary growth, growth localized at a growing edge, it is common to measure separation along axes which vary with the developmental stage of the organism in some natural fashion. Moss and Salentijn (1970), believing the growth of the human mandible to be of this form, measure distances along a certain logarithmic spiral which includes three mandibular foramina. Raup (1966) models the snail shell by a basic plane section simultaneously rotating about an axis, moving along that axis, and increasing in scale. Shiells (1965) measures a growing shell along its circumference.

A second possibility is the provenance of a specialized multidirectional coordinate system with symmetry properties. Needham (1950) used the principal body axis and the abdominal segmental boundaries normal to it to derive a true two-dimensional growth-gradient for the crab Pinnotheres pisum. Figure V-6 summarizes his quantification in a remarkable Cartesian transformation grid much clearer than a contoured plot of directional k's could be. Such an analysis should not be confused with similar analyses of two-dimensional gradient fields in one direction only, as in Needham (1964: fig. 4.6), or of one-dimensional gradient fields followed over successive stages, as in Needham (1937), Blackith, Davies, and Moy (1963), or Brown and Davies (1972). It is very difficult to combine gradients and grid in the same diagram. The illustrations of Ambystoma larvae in Richards and Riley (1937), for instance, manifest far too many straight lines to suggest any Thompsonian homogeneity of deformation.

Akin to symmetry in space is constancy over time: systems in a

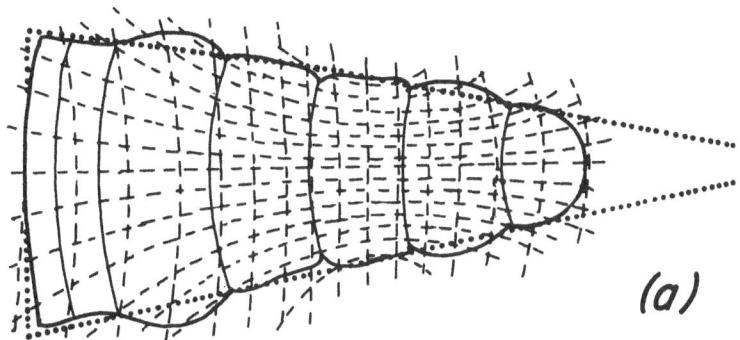

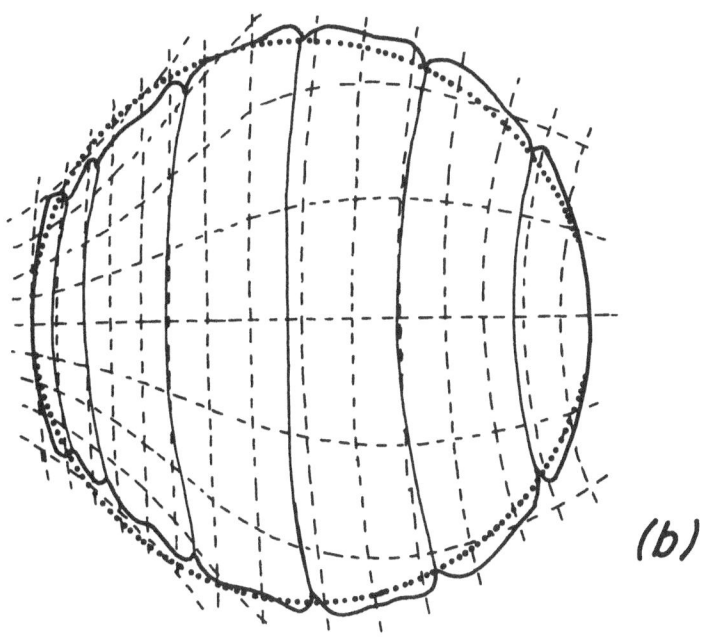

Fig. V-6.--Cartesian coordinate transformations between the
adult male (A) and the adult female (B) of <u>Pinnotheres</u> <u>pisum</u>, the
pea-crab. Each grid is the trace of a grid squared upon the other
form. After Needham (1950).

"steady state" of growth. For instance, there is a considerable
botanical literature on the subject of shoot and root meristems,
centers of morphogenetic patterning which lie in invariant ·spatial
relation to the steadily differentiating mass of tissue produced in
their past. Schüepp (1952) places upon sections of meristems coordinate curves which are trajectories, relative to the growth center, of particular cells as the continued growth of the meristem
pushes it away from them. In the resulting diagrams, any two regions each bounded by a cross-cut of those trajectories at points
of constant age are true Thompson transforms of each other, for
they correspond point for point inside. In such stable growth patterns, time is an ignorable coordinate. Any future snapshot is
identical to the present snapshot with the points systematically
relabelled. Certain spacings along those trajectories in any section then correspond to specific local growth rates across sections. '
A broad geometric theory of meristematic activity making extensive
use of this particular sort of symmetry may be found in Schüepp
(1966).

Both these extensions of gradient analysis evade Thompson's
demand that the dependent variable should be form itself. A method
appropriate for study of segment series cannot be modified to analyze
vertebrates: the higher class has no natural segment boundaries
along·which to align the vector separations worth measuring. These
boundaries are perpendicular to a growth-gradient along a principal
body axis, but in general shape changes there is no such axis (away
from the midline) for correlating directions at finite distance.
Distances measured today along a straight-edge will correspond to
arcs more or less curved tomorrow (Thompson, 1961:320), but the
proponents of growth-gradient analysis do not instruct us in which
arcs to use. Somehow the features of form--the bulges and bends,
convexities and concavities and protuberances--are lost in the reduction to quantity. This will not do: there is not enough geometry left.

C. Simulations

In this section I list three analyses unknown to each other
which I find to share a common spirit closer to Thompson's. Each
one investigates growth-rates in the small throughout an organ, and
then discusses the extent to which the separate local changes cohere
in the global change of form actually observed. The data are never

quite capable of explicitly supporting the leap to larger patterns,
creating a logical gap which is bridged by mathematical assumptions
that simulate the final change of form.

Green (1965) studies the patterns of surface growth which might
account for a particular observed invariance of shape in three dimensions. His subject is the apical tip of Nitella, which maintains
its shape--a hemisphere surmounting a cylinder--as it incorporates
new plasm and grows upward. It is a stubborn mathematical fact that
the preservation of shape does not determine a unique local growth
rate field. Anisotropy is optional; it need merely be properly coordinated with the change of directional gradients over a curving
surface. For radially symmetric systems such as the Nitella tip,
alternative growth models can all be expressed as functions of
meridional distance from the apex. Different functions correspond
to quite different local mechanisms, for while isotropic growth
suggests scalar morphogenetic fields, anisotropic growth requires a
histological asymmetry grounded in some cytoplasmic texture. A certain mathematical elegance is associated with this analysis, as form
invariance leads to a differential equation directly governing the
rate function. The calculus is developed further by da Riva Ricci
and Kendrick (1972). To settle the question of isotropy Green pro-
ceeded to measure actual rates of increase of separation between
points marked with microspheres on the growing hemispherical surface. He finds, in fact, anisotropies which reverse their sense
over cell age. The origin of directionality in the known structures
of plant cell walls is reviewed in Green (1969). Such a strategy,
despite its elegance, cannot be expected to generalize to systems
lacking the peculiar symmetries of this organ, radial symmetry and
invariance of form over time, as the necessary differential equation
can no longer be produced.

The anatomist D. H. Enlow has long studied the details of human
bone growth. His findings regarding the craniofacial region are
collected in Enlow (1968, 1975). Bone grows generally by "remodelling," unending deposition and resorption. These alternatives can be
discriminated in photomicrographs. Enlow has carefully examined the
nature of growth on every surface, across every tuberosity, of the
normal craniofacies. He is thereby able to specify, for any particular anatomical form of interest, what local growth processes, taken
in concert, are responsible for the observed changes of form and
relative position. There are certain recurrent themes, for instance

that V-shaped forms generally grow by deposition on the inside of the V, resorption on the outside, and thus displace themselves as they grow.

From his analysis, he concludes that form moves through bony tissue, slowly changing under the impulse of external correlations and alignments. There is an insight here which goes beyond Thompson. Growth may remove material as easily as it deposits. Shape is in part negative, adjustment to space outside the material boundary. In fact, in a growing bone maintaining a shape which is not convex, certain surfaces are always necessarily shrinking, undergoing "decretionary" growth. In addition, within the face as a whole, bones are moved passively by abutting on other bones themselves growing. In principle such a system for analyzing growth should account for all changes of bone shape, abnormal as well as normal, in terms of the two basic processes, deposition and resorption, all over the growing surface. Unfortunately, the quality of present numerical data is insufficient to support quantitative analysis systematizing the detailed qualitative and directional analysis. We do not know how to talk clearly about widely spaced simultaneous remodeling processes which together very nearly preserve a functional form. If a geometric formalism could be invented usefully to represent the elegant osteology of this method, analysis could advance rapidly, and perhaps I might no longer consider this work a "compromise" with Thompson.

My last example of simulation is an intensive study of one phase of newt neurulation (Jacobson and Gordon, 1976). The data are a scalar field of cell size decrease from stage 13 to stage 15 of the embryo, together with measures of the notochord's exogenous extension during that period. The analysis produces, according to a mathematical model of tissue shear, a predicted shape change for the whole neural plate. In the form of a summary Cartesian transformation grid, Fig. V-7, this simulation may be compared with an observed grid previously published; it is a fair replica. The most interesting feature of the simulation is the isolation of anisotropy. The separate cells of stage 13 shrink in cross-section individually and isotropically, whereas the extension of the notochord is highly directional. The authors put forward a rather complicated model for the propagation of this shear through the bulk of the tissue by rearrangement of cell-to-cell connections. At present this phase of the simulation has an irreducibly stochastic component,

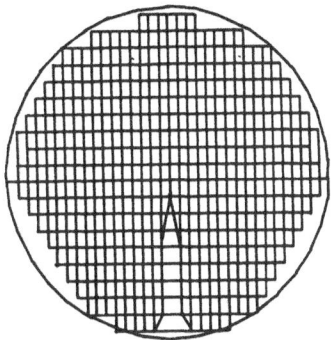

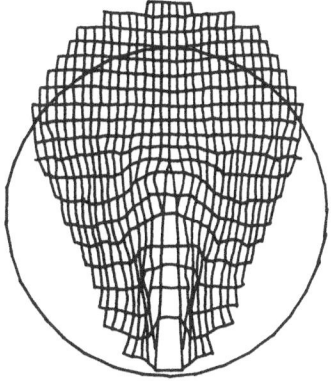

Fig. V-7.--Cartesian transformation grid for the simulation
of neural plate development. The
wedge in lower center is the developing notochord. After Jacobson and Gordon (1976).

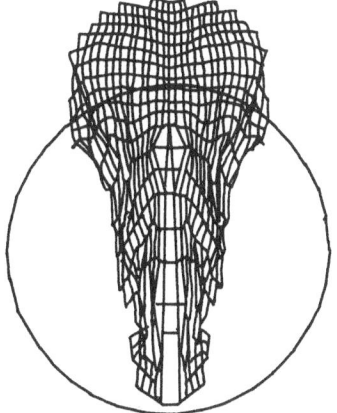

which I trust can be removed by further mathematical analysis. But
it is not necessary to believe in the reality of this mechanism to
appreciate its elegance and parsimony. In differential growth
analysis over areas, one consequence of anisotropy is an increase
in the number of essential numerical parameters from one to three
(two growth rates, or areal growth rate and anisotropy, together
with orientation of the principal axis). The Jacobson-Gordon
model shows that over most of a tissue one can make do with but the
scalar representation if the more complex specification of anisotropy is present as a driving force abutting the simpler system.
Their conclusion is quite in the spirit of Thompson: "The joint
operation of two physical forces is necessary and sufficient to effect this transformation" (Jacobson and Gordon, 1976:191).

None of these simulations presently are capable of mathematical generalization to transformation grids in their full complexity. Each contains at least one good idea which ought to be present in any more inclusive scheme. From Green, the nugget is the treatment of surface elements; from Enlow, the idea of simultaneous positive and negative growth; from Jacobson and Gordon, the modeling of anisotropy as viscous response to exogenous shear. All are suggestive, but none are yet generally satisfactory.

D. Other Morphometric Schemes

Vector displacements. A body of work directly relevant to our theme arises from the problems faced by the practicing orthodontist. He is mainly in the business of manipulating the shape represented by a sagittal cephalogram, an x-ray of the head from the side. The mature shape he predicts may be thought of as a geometric transformation applied to the growing form at any stage, a transformation approximating somehow the mean changes in an appropriate reference population followed over time. By another transform, any particular face is more or less distinguished from an aesthetically ideal shape; to characterize the transformation among these two is to provide a hint of how the orthodontist might restore grace and balance.

As I noted in chapter iii, in cephalometrics two shapes are compared either by contrasting their separate interlandmark measures or by superimposition of their drawings. In the latter case, the pair of shapes are geometrically related by the vector displacements from image to image of all the landmarks the investigator cares to follow. A Cartesian grid may be used to summarize all these displacements if it is somehow aligned to express the requisite registration and orientation. Diverse investigators--de Coster (1939), Izard (1950), Moorrees and Lebret (1962, 1975)--have established "standards" for such a mesh in diverse ways. Krogman and Sassouni (1956:251,283) demonstrate that the different methods yield grids rather different in appearance and in quantitative details. The basic difficulty is that the techniques locate landmarks but not curves connecting the landmarks. It is then ambiguous how the grid lines are to be curved in the great spaces between, even though it is this curvature which embodies the visual impact of any mesh analysis. Also, each method aligns the basic "graph paper"--the grid for the normal image--differently upon that image.

The manner in which Moorrees constructs transformation grids shows peculiarities typical of this approach. The nasion is assigned two fixed Cartesian coordinates, the sella turcica one, the

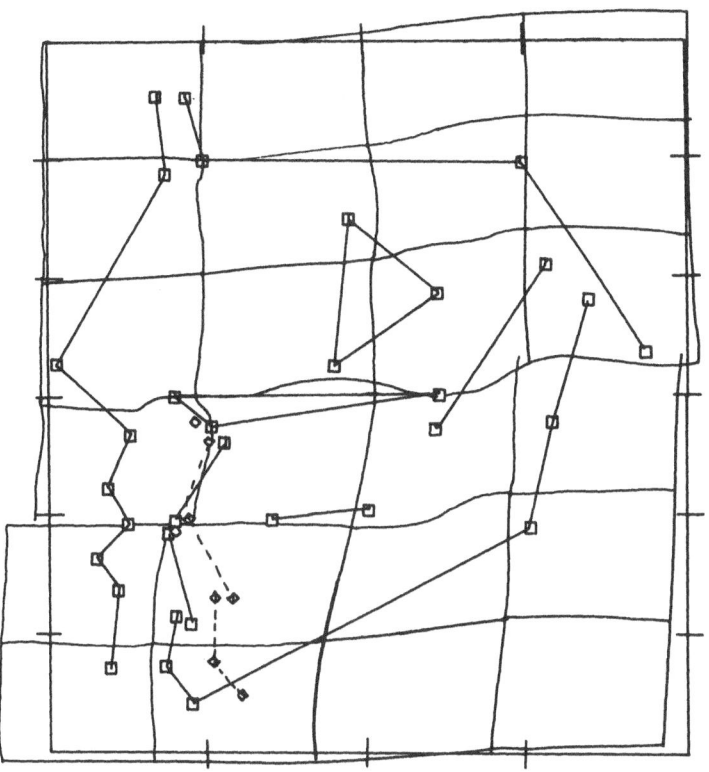

Fig. V-8.--Cartesian pseudogrid for a normalized cephalogram
indicating "marked mandibular prognathism and cephalad position of
sella turcica, basion, articulare, and gonion." The facial profile is to the left. After Moorrees, et al. (1975).

anterior nasal spine one. The horizontal and vertical axes of the
grid are scaled separately and oriented according to "natural head
position." The actual grid, which may be drawn by hand or by computer, is figured separately for the upper and the lower face, with
a discontinuity along the line of occlusion. Having produced a
sort of transformation grid, as in Fig. V-8, Moorrees and his col-
leagues do not really know what to do with it. In fact, the complex registration has made it statistically intractable. The grid
is but a picture of a certain idiosyncratically standardized finite
collection of vector displacements. The original biological form
has been altered in the engridment, because of the separate scalings

in two directions. The grid as drawn is not the transformation to
the observed form from normative configuration or past form, because a new anisotropy has been imposed: the transformation from
x-ray to diagram is affine, not Euclidean, and should have been
fitted by a weighted average of shears throughout the form. Also
by virtue of the registration, certain points of the grid are intrinsically less variable than others, quite without reference to
the complex ontogeny of the cranium and face.

The registered grids have one great temporary advantage. They
remove just enough of the "floating" quality of Thompson's general
framework to make possible a multivariate statistical analysis,
however flawed. Vectors and displacements have population averages
and standard deviations which can be presented for use in diagnosis
(cf. Merow, 1975). The "profile" so noted does not, for all the
reasons noted in chapter iii, represent a true shape comparison between normal and observed or over successive stages of ontogeny.
We cannot analyze Thompson's grids in general by studying the displacements of all the grid intersections from one fixed intersec-
tion--to do so would be to lose all the variation of the fixed intersection, to misapprehend all the factors of systematic variation
in which that fixed point is implicated. The variables pictured
here are simply not what we need. The method of transformations
has no role for fixed points or orientations; it is couched in
terms of how points grow apart from each other, not from an arbitrary center.

Multivariate morphometrics. One might imagine that a method
for measuring shape could be adapted easily to measure shape change,
and would thereby automatically make a contribution to the study of
transformations. This is simply not true. In terms of measures of
shape, however subtle, shape change can only be an abstract vector
of differences, one for each measure. This series is in no way the
unique mathematical object which Thompson steadfastly envisioned.
The landmarks and positions suitable for characterizing shape are
likewise not necessarily suited to the characterization of change--
we may need to know where everything is going, the relative motions
of intercalated points as well as landmarks.

Nor are the subtleties of biological correlation between separate changes in parts, which is the wellspring of Thompson's model,
expressed adequately in variance-covariance matrices. All multivariate techniques, however resourceful, are restricted to linear

relations among the variates; but the reality of variation in form is often functional and far from linear. Pearsonian correlations between characters, which underlie the canonical styles of variation a multivariate analysis finds, can only hint at the nonlinear associations underlying the observed transformations of form.

Of all those working in the new multivariate morphometrics, the only one, to my knowledge, who acknowledges these limits is Oxnard. In his collected thoughts on the analysis of form (Oxnard, 1973), he takes pains to verify statistical insights by comparison with simple biomechanical demonstrations about the observed regimes of variation. All others--for reviews see Blackith and Reyment (1971), Corruccini (1975), Gould and Johnston (1972), and Kowalski (1972)--identify variation on empirically derived canonical linear combinations with the biological mechanisms producing the variation. This formalism of abstract measurement spaces could not be more divergent from the geometric insights of Thompson's scheme. The multivariate techniques capture shape in a way which necessarily misrepresents shape change. The spokesmen for this school tend to invoke Thompson's name (Gould and Johnston, for instance, describe his vision as "multivariate"), but they are only studying variation of their indices, not of the underlying shapes.

At the conclusion of my survey to 1977 of the literature on Cartesian transformations, I find no improvement from within morphometrics or without, no methodological advance for particular styles of data, that is comparable in stature with Thompson's original method. Six decades after its publication the method still resists quantification except in special cases. It remains a fascinating puzzle for biostatistics and mathematical biology, endlessly suggestive and promising, but much more difficult than it was supposed to be. There is yet no progress. Anyone trying to make new headway must begin to build, as I do, exactly where Thompson left off.

CHAPTER SIX. THE METHOD OF BIORTHOGONAL GRIDS

Like several of the authors reviewed, I too have been tantalized
enough by the recalcitrant elegance of the Thompson problem to attempt
a fresh mathematical unfolding.

A. Representation of Affine Transformations

My researches began with an aspect of the Thompson method not
previously noted: that the features of a grid as we apprehend them
depend capriciously on the grid with which we begin. Consider, for
example, the transformation of a square into a 60° rhombus, drawn
twice in Fig. VI-1. The upper engridment would be described as a
vertical shrink by the factor $\sqrt{3}/2$, the horizontal unchanged, followed by thirty degrees of shear. In the lower engridment there is
instead no shear, but one axis shrinks by $\sqrt{2}/2$ while the other expands
by $\sqrt{6}/2$. Squash or shear? Either. There is no contradiction here.
The description does indeed depend on the choice of starting grid,
just as the elements of a matrix representing a linear transformation vary when the basis is altered.

I suggest that the lower of the two representations in Fig. VI-1
is to be preferred. For affine transformations, those which take
parallel lines into parallel lines, such a pair of axes is guaranteed.
There will always exist a pair of directions, the <u>principal axes</u> of

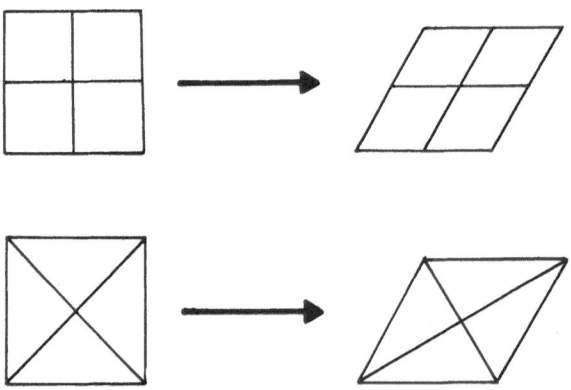

Fig. VI-1.--Two reports of the same affine transformation.
(Top) Axes do not change in length, but their angle is altered
from 90° to 60°. (Bottom) One axis shrinks, the other expands;
their angle remains at 90°.

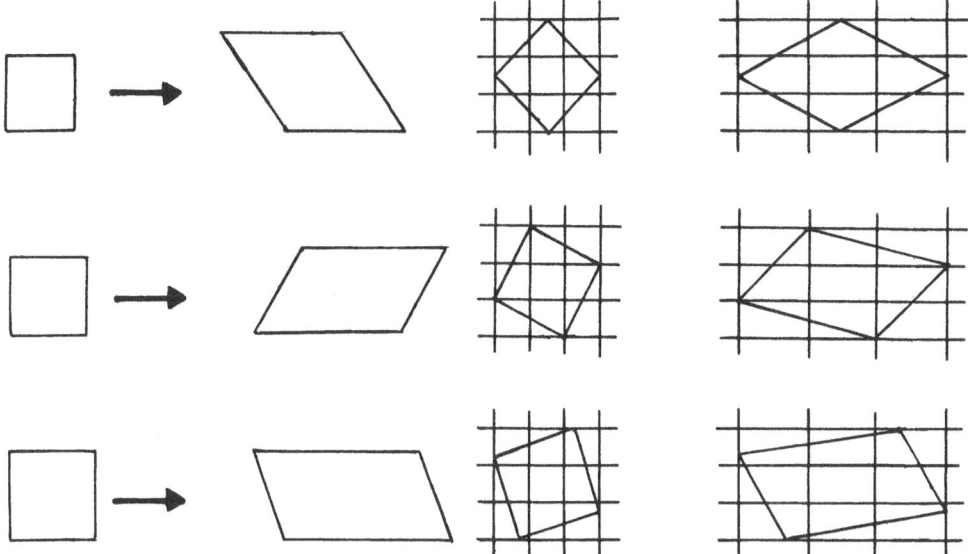

Fig. VI-2.--The same shape change, a 2:1 stretch, attached to a square in three different ways. As the orientation of the axes is altered, the visual impact of the affine transformation varies in a complex manner.

the transform, which are perpendicular both before and after transformation, and with respect to which the observed shape change is described by a pair of <u>dilatations</u> (length-multiplications, stretch or shrink), one along each axis. A third parameter fixes the direction of these axes in the image.

In Thompson's original technique, the starting grid is aligned with some feature of the anatomy, the single shape. I would align instead with the shape change, regardless of the initial form, which just cuts out a piece of grid after the fashion of a cookie-cutter. In Fig. VI-2, for example, is shown a single transformation, with dilatations of 2 and 1, repeated three times. As the orientation of the initial square is varied, the reports of side-stretches and shear, and the visual impressions of the left-hand engridments, change in a complex manner; but the report using principal axes, on the right, shows, of course, two parameters constant, only one changing.

Using these axes we can recognize iterations and inverses of a fixed transform with ease. The transformation shown in Fig. VI-3 is the result of applying that of Fig. VI-1 twice in succession.

This is not at all obvious in the Thompson engridment, but it is
readily recognized using the representation by principal axes, for
the axes of Fig. VI-3 are the same as those of Fig. VI-1, and the
dilatations of the latter (.5, 1.5) are just the squares of the
dilatations of the former. Now the inverse of the transformation
from A to B, where A, B are any two homologous shapes, is of course
the transform from B to A, and our description of a transform should
be clearly opposite to the description of its inverse. This does
not obtain under arbitrary orientation of source grids. For instance,
Fig. VI-4(a) depicts a twofold vertical stretch combined with a 45°
shear. Its inverse in the Thompson formulation is gotten by switch-
ing the square-gridded coordinates to the other image, as I have done
in (b). The inverse transformation then may be seen to be a twofold
vertical diminution followed by a shear of 63°. This certainly fails
to invoke the inverse nature of the transformations. We see from the
principal axis representations (c,d), however, that the one has dilatations .7, 2.85, the other $.7^{-1}$, 2.85^{-1}, and that the axes are the
same.

Transforms like these, affine in the large, occasionally describe real data, as whenever fossiliferous rocks are strained by
geological processes carrying embedded fossils along. When a few
finite distances and directions can be paired between distorted and
undistorted forms, the methods of ordinary least-squares fitting can
be made to reconstruct analytically a best estimate of the strain
tensor and its principal directions. See Sanderson (1977) for an algebraic exposition of the general method, which geologists have used
for decades under the name of "analysis of strain rosettes." Further
special tricks exist to be invoked in suitable special cases. Elliott (1970) explains how to estimate the deformation matrix for
randomly oriented ellipses about whose initial distribution something
is known. Hirsinger (1976) presents a Fourier method for the reconstruction of strain from two forms, differently oriented, of the same
(irregular) shape. Tan (1973) elegantly solves the estimation problem for logarithmic spirals, as of ammonite outlines, whose self-
similarity makes moot the absence of landmarks or orientation.

B. General Lines of Growth and Biorthogonal Grids

In all these methods, the presumption of affineness in the large
enforces a rigorous interdependence among dilatations and angles at
finite distance which is untenable in general. For non-uniform dis-

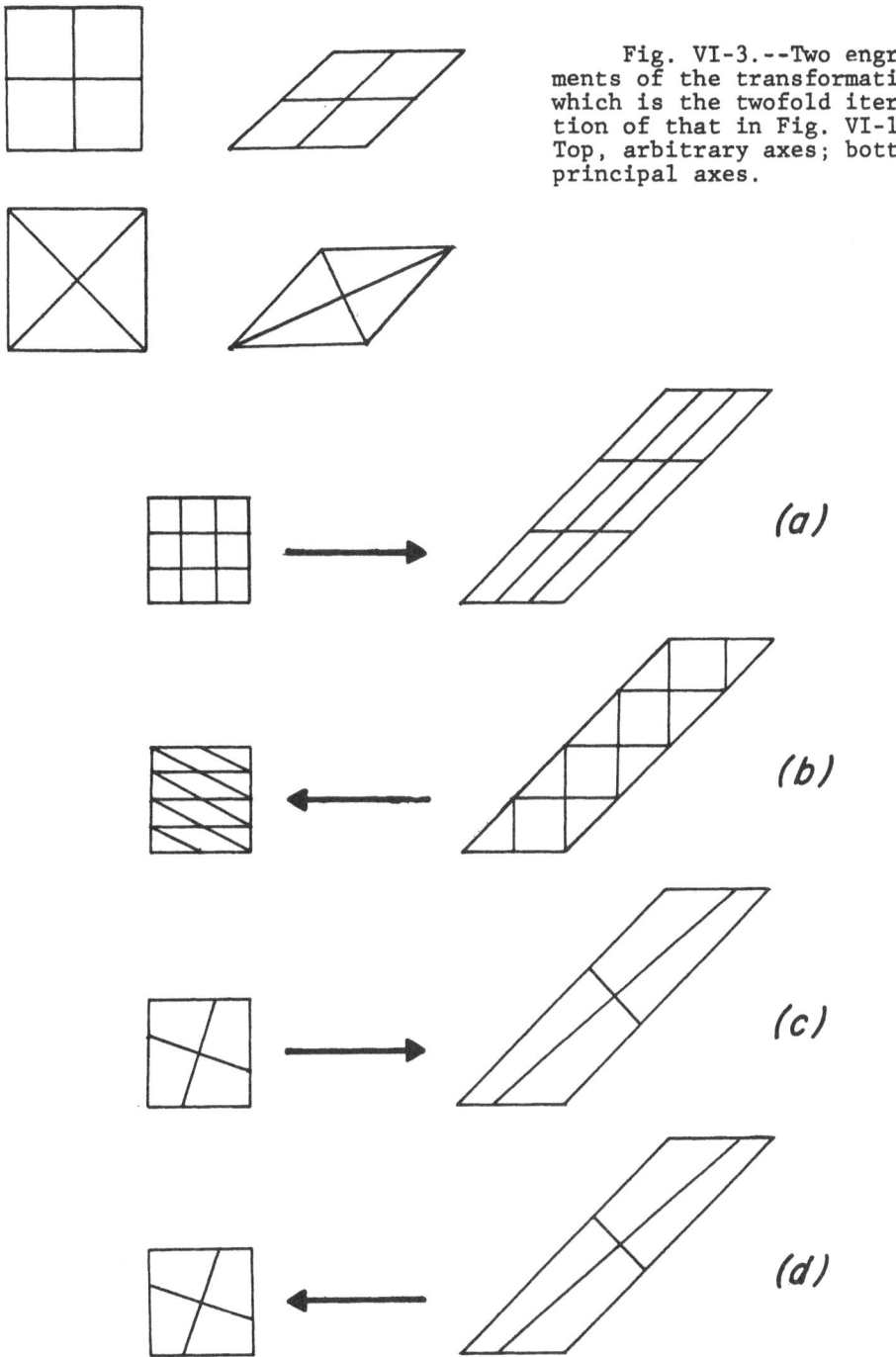

Fig. VI-3.--Two engridments of the transformation which is the twofold iteration of that in Fig. VI-1. Top, arbitrary axes; bottom, principal axes.

(a)

(b)

(c)

(d)

Fig. VI-4.--Two engridments of a transformation and its inverse. (a,b) arbitrary axes; (c,d) principal axes.

tortions, the several regions interrelate only according to constraints of continuity, and the search for globally perpendicular line fields fails. For instance, we might explore alternate realizations of the simple projective transformation drawn in Fig. VI-5. With the axes parallel to the sides of the square, frame (a), we see one vanishing point, and all intersections on the midline remain at 90°. With axes along the diagonals of the square, frame (b), we see two vanishing points instead of one, and angles are distorted even on the center line. The 90° grid angles remain 90° only on a certain curve that is not a straight line at all. In more general transforms than the affine we cannot find two families of parallel straight lines at 90° which transform into two families of parallel lines at

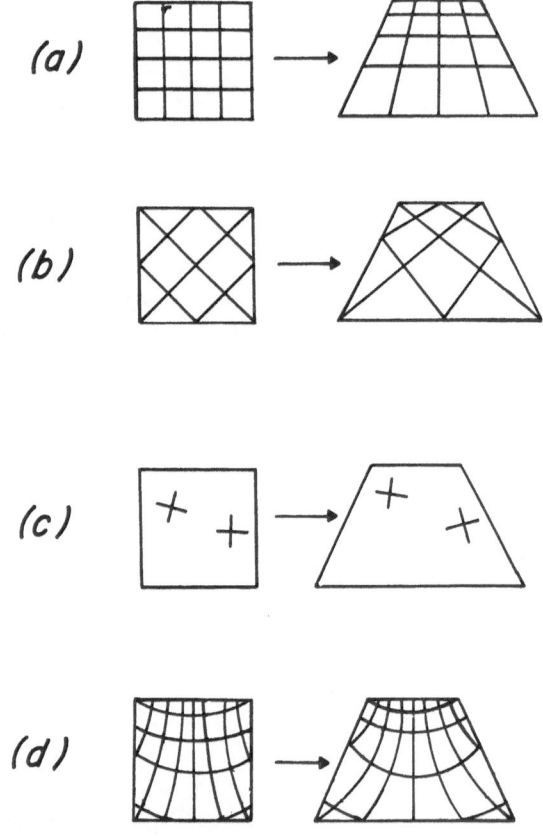

Fig. VI-5.--(a,b) A projection from window to its shadow, in two arbitrary sets of axes.   (c) Through each point of the correspondence there is one pair of lines which start and finish at 90° to each other.   (d) We represent the transform by the integral curves of these local perpendiculars.

90° in the new image. It might occur to us to search for curved lines which satisfy this property. After some experimentation with the formulas for projection, and after much trial and error, we might sketch the system of curves orthogonal at start and finish of this projection map. My drawing is the bottom pair, Fig. VI-5(d). Such a figure seems at first glance to be something from a proof in a geometry text, quite unrelated to the biological problem at hand. But the suspicious perfection of these curves is only an accidental implication of the highly idealized projection transform itself. It is not the curves themselves which embody the growth, but rather small regions around their intersections. Through every point of the square pass two perpendicular lines mapped into lines perpendicular in the trapezoid, as in frame (c). The orientation of these lines rotates from point to point of the shapes. The smooth curves of frame (d) are each tangent everywhere to the appropriate local principal axis, so any two curves are perpendicular wherever they intersect, in either image. We cannot draw curves through all points, so we select from the two families to suggest their changing directions by a curvilinear grid. We thereby arrive at a distorted grid in which the curvature of the curves depicts the movement of the principal axes, also called "canonical axes," across the diagram.

The curves do not have a biological reality as loci. In particular, their spacing on the one side or the other is an arbitrary decision of the draftsman. They are but threads along which to string little right angles, perpendicular intersections, tangents and normals, which do describe the local biological reality. Let us consider one infinitesimal neighborhood in a biological tissue and its canonical axes in the original and in the transformed shape, as in Fig. VI-5(d). With respect to the canonical axes, the observed growth or expansion is symmetric. In the small, little square boxes of sides aligned with these axes, and only these, grow into little rectangular boxes similarly aligned. All other squares grow into oblique parallelograms instead, as in Fig. VI-1. Along these special axes, as in the affine case, the transform is described completely by two local dilatations, which now vary in magnitude all the while the axes change direction. If the dilatations are 2 and 1.5, for example, a little box of 2 cells by 2 aligned on the local canonical axes expands to a little box 4 cells by 3 along the corresponding canonical axes in the other shape. If these axes are rotated on the tissue it is because of growth in other parts of the beast. The little box, from its infinitesimal point of view, is just growing at two rates in two directions, and it doesn't know anything about rotations.

Enough of the mathematics has now been explained to allow of stating a postulate: the lines of growth are biologically invariant. If they are fixed, as by India ink, as hachures in a developing tissue, they transform into the corresponding lines of growth at every future stage. In local terms: any infinitesimal 90° angle which remains at 90° through any finite transform (i.e. which is such that growth is symmetric around it, as in Fig. VI-5(c)) remains at 90° through all intermediate stages of the transform and will stay at 90° for the foreseeable future. Any India-ink curve which is tangent to a side of these canonical perpendiculars for any finite transformation will be tangent to such a side for all finite transformations past and future. The postulate in effect declares: nothing really bends. All apparent curving and rotating is produced by differential growth along axes fixed unbendingly at 90°. Axes rotate only when they are pushed.

Such a postulate clearly distinguishes the mathematical model from the actual processes underlying shape change. Throughout this essay I mean to imply by the term "growth" simple adjustment of outline, for I have excluded all other sorts of information from the data base. "Lines of growth" are descriptive mathematical artifacts, not anatomical features and certainly not causal agents. In ontogeny, shape change is the outcome of many different fundamental biological processes: cell replication with differential division, change of cell shape, morphogenetic movements, cell death. These are responsive to continuous gradients, discrete cellular effects, and functional needs and constraints. The mathematical model presented here attends to none of these known and hypothesized processes, but only to the resulting geometry. In all comparisons which are not ontogenetic that we might study, individual with norm, siblings, endpoints of evolutionary lines, we are even farther from actual processes, as we deal with subtle traces of ontogenetic control systems. It is necessary to assume that points of homologous pairs arise from homologous Anlagen and differ only in the quantitative details of the dilatations which brought them to the measured locations.

This stance of stubborn geometric reductionism is necessary in view of the great logical difficulty of comparing whole shapes in the absence of a complete material history. The grids of Fig. VI-1 are but alternate graphics for the selfsame biological comparison, and the choice between them is a matter of analytical convenience. I am recommending the lower frame, of course. In Fig. VI-2, however, are shown three different biological transformations. They all have the same mathematical form but have different lines of growth, different attachments to the living tissue. Actual growth might combine these in unpredictable sequence in between the observed sections. The lines still represent the space of possible models from which we draw, and morphogenetic movements, which the model cannot recognize, simply move the lines to a new position which compromises with the unconformity. In this way we systematically advance the task at hand: to model the shape change in a consistent geometrical way that allows for comparisons and extrapolations. We must somehow compare whole shapes in the absence of a complete material history. When the postulate of invariance of the lines of growth conflicts with obser-vation, we must go with the data: we say that the growth-gradients have evidently changed. Otherwise we make the simplest possible as-sumption, that things which have appeared unchanging (the lines of growth, intersecting consistently at 90°) will continue so.

To extract information about growth, we compare shapes of match-ing little boxes in a grid and its transformed image. In any comput-able grid at finite spacing, these boxes will be not quite rectan-gles; in the limit of infinitesimals, they are exact rectangles. For the projection transform, Fig. VI-5(d), we construe the splaying of the originally square vertical sides as a consequence of excess of horizontal dilatation over vertical. The sides have been pushed out; we feel the axes shoring themselves against their perpendiculars in order to exert the appropriate stress. (The shears of Thompson's engridments contrariwise lead one to a metaphor of stresses exter-nally imposed, a vision less useful.) We read dilatations directly from the diagram-pair as ratios of grid spacings in the two images. The convergence or divergence of curves for one shape separately has no particular meaning.

This method is perfectly general. Through almost every point of a differentiable transformation pass just two differentials which are at 90° both before and after transformation. The integral curves of these differentials form a grid whose intersections are at 90° in both images. These are called the _biorthogonal grids_ for the trans-formation, for there are two of them, one in each image, corresponding curve for curve, intersection for intersection. All these assertions are proved in Technical Note 1 below. We quantify shape change by extracting the two dilatations at every point of either shape, measured along the local canonical axes. Thompson's fundamental error was the construction of diagram pairs which were unsymmetrically specified: rectangular grid on one side, unrestricted grid on the other. In view of the symmetry of all growth in the small, the appropriate grids for any two shapes have the same formal property, that all grid intersections are at 90°. <u>This characterizes the bioorthogonal pair uniquely</u>. We have thereby produced a canonical coordinate system which reduces all change of shape to gradients of differential directional growth, without shear. As with the affine transforms, Fig. VI-2, we have specified a coordinate system for the change itself. It becomes possible to measure shape change, then, without measuring shape at all. In fact, the measurement of particular shapes is subsumed under this rubric as the transformation from that shape to some predefined norm.

The method of biorthogonal grids cuts through several of the methodological difficulties common to its predecessors. First, with passive rotation removed, growth is represented by those two dilatations with orientation, in effect a symmetric tensor field over one whole image which can be compared to others on the same image. It is registered everywhere. There is therefore no need to register the images in advance. The only privileged points are the landmarks which we use to calculate the one-to-one correspondence: they happen to align the grid but are not discernible as special points within it. In principle, the results of any of the existing schemes for extraction of quantity--growth-gradients, vector displacements, multivariate morphometrics--can be reconstructed from a collection of biorthogonal grids, and their biases owing to registration and orientation can even be analytically derived. Second, the technique prescribes explicitly the distances, directions, and gradients that are worth examining. This is managed by abstracting away two particular empirical possibilities, the "moving forms" of Enlow and the tears and morphogenetic movements of Moorrees or Jacobson and Gordon. In the absence of these more precise models of actual change, in which landmarks do not stay landmarks and points arbitrarily close together end up far apart, the proper mathematical model is that of a diffeomorphism, from which the dilatations and principal axes follow rigorously. This is precisely Thompson's own assumption, and so this quantification answers precisely to the needs of his method.

Singly, orthogonal coordinate systems of this sort are used in diverse other sciences, wherever Cartesian tensors are applied. In cartography the existence proof of the axes described here is known as Tissot's Theorem (Richardus and Adler, 1972: sec. 3.5). The axes are the principal axes of little ellipses drawn on a map which represent the linear distortion of map distance as a function of direction on the surface, e.g. the geoid, being mapped. In differential geometry the lines of curvature of a surface, which have as tangents the directions of extremal curvature for normal sections of the surface, form an orthogonal net of just this sort (Kreyszig, 1968:177). In continuum mechanics, stress, strain, and various other tensor fields are commonly described by reference to their principal axes. In biomechanics, it is an old hypothesis that trabeculae of bone are laid down along integral curves of the principal axes of stress (Thompson himself discussed this) and a newer one that the fibers of the heart muscle are laid down so as well (Peskin, 1975). Except for my biorthogonal grids, I know of no application in which orthogonal coordinate systems are derived from empirical data rather than deduced from mathematical formulas a priori.

C. Summarizing the Grids

By reference to the biorthogonal grid the summarizing of a transformation becomes particularly simple. Since all angles within the grid are unchanging by construction, such change of form as has occurred is a matter of changes of relative curve-spacing only, of differential _dilatation_ (stretch) from point to point in the two local principal directions. We report the ratios of corresponding segments in the two grids, and refer to this as the "relative growth" from one to the other. Along any integral curve these ratios increase or decrease in growth-gradients _sensu_ Huxley (1932).

In practice, the grids appear to be aligned in very sensible ways. Consider, for instance, the transform in Fig. VI-6(a), the growth of an abstract "foot." The input data consist of two sets of four corners expressed in Cartesian coordinates of separate origins, orientations, and scales. Filling the forms are two meshes, corresponding point for point between the images, which summarize the homologies of the interiors. (The algorithm defining their correspondence is set forth in Technical Note 2.)

These meshes might be connected by straight lines in the beginning form, smooth curves in the ending form, resulting in a Cartesian grid pair after the style of Thompson. But I shall proceed in-

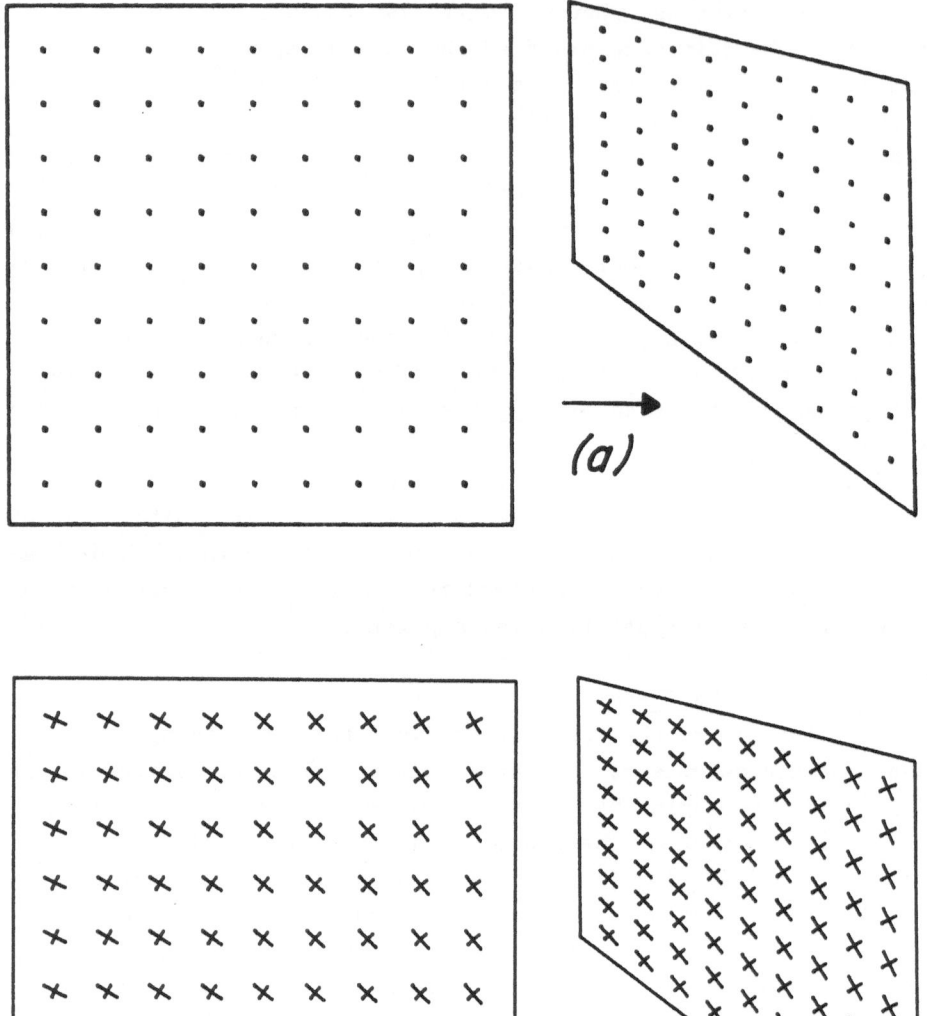

Fig. VI-6.--Biorthogonal analysis of an artificial example.
(a) The homologous outlines and a mesh representing the optimally
smooth interpolation between interiors, which is here bilinear.
(b) The local principal strains, homologous directions at 90° in
both images, sampled on the mesh of (a) by numerical differentia-
tions of the map shown in (a).

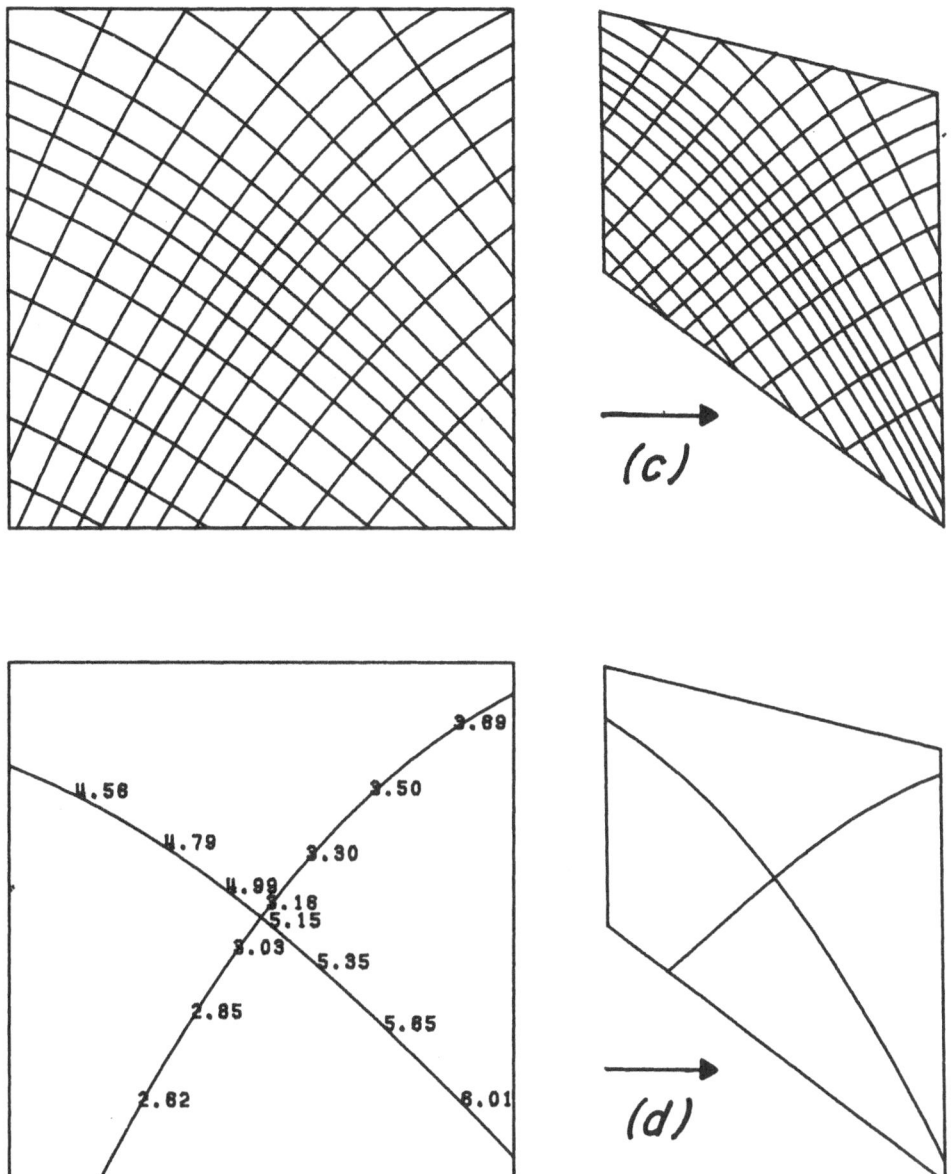

Fig. VI-6.--Biorthogonal analysis of an artificial example, continued. (c) A grid of integral curves for the tensor field of principal directions which was sampled in the preceding frame. (d) Two
integral curves, and the dilatations from square to quadrilateral of
corresponding segments along them. The scale of the right-hand form
has been reduced for drawing.

stead using the field of principal directions, Fig. VI-6(b). Each
little cross resides at one of the original grid points and expresses
the strain field locally in terms of orthogonal dilatations along its
arms in the ratio of the lengths of those arms. In the right-hand
image, the crosses are to a consistent scale throughout, thus embodying also the ratios of dilatations from point to point. It may
be seen how these little x's rotate and stretch gently between sectors of the figures.

A rather more legible presentation of this same geometric object,
a symmetric tensor field, is by way of some of its integral curves
assembled into a biorthogonal grid pair, as in Fig. VI-6(c). If this
figure were superimposed over the preceding, it would be seen that
each curve is parallel to one arm or the other of the little crosses
at every point through which it passes. These curves, then, provide
a geometric linkage for finitely separated deformations (expressed
by those crosses) whose definition is purely local, in terms of pointwise affine derivatives.

In Fig. VI-6(d) I extract two particular curves from this grid,
curves not bearing special optimality properties but rather typical
of the grid as a whole. The decimal numbers spaced upon these curves
represent the dilatations along them, ratios of distances in the
quadrilateral to the homologous distances within the square. (The
scale of the lower form was reduced before drawing, so that the
printed dilatations all appear larger by the factor 6.0 than the corresponding ratios of lengths on the page.) The gradients of these
quantities are equivalent to the gradients of ratios of corresponding segments between intersections of the richly drawn grids, Fig.
VI-6(c), from which they differ only by that constant factor of scale.

A report of distances and angles among the corners in both figures, while encapsulating the actual shape change (by wholly embodying the data), would be tedious and would not isolate "factors" in
any useful sense. The biorthogonal analysis is much more suggestive.
The major axis of growth is slightly bowed near one diagonal with
dilatations graded from 4.5 at one end to 6.0 at the other. Perpendicular to this, and also nearly aligned with a diagonal of the original square, is a minor axis with dilatations graded from 2.6 to
3.7. The lower right corner has grown away from the center most
rapidly, the lower left least rapidly. The remaining axes and dilatations intergrade nicely. By means of the biorthogonal display we
have been able to summarize this entire transform in but two main

gradient-bearing curves. There is no need to report changes of angle--there is only change of scale, continuously varying in alignment and magnitude. The change is encapsulated in the derived coordinate system, together with its orientation upon the original form; the original configuration of landmarks, their relative distances and angles, is quite accidental and has not distorted our quantification of the change itself.

The method of biorthogonal grids has reduced change in shape to differential changes in size, and measures shape change without confounding by shape at all.

Technical Note 1. Existence and Form of Biorthogonal Grids

In this note I demonstrate formally the existence of biorthogonal grids for arbitrary Thompson transformations.

Let $u(x,y)$ be the x-coordinate of the image point corresponding to original point (x,y), and $v(x,y)$ likewise the y-coordinate. The Thompson transform is then the map $(x,y) \rightarrow (u(x,y), v(x,y))$ taking a region of the (x,y)-plane onto a region of the (u,v)-plane. Assume the existence of derivatives $u_1 = \partial u/\partial x$, $u_2 = \partial u/\partial y$, $u_{11} = \partial^2 u/\partial x^2$, $u_{12} = \partial^2 u/\partial x \partial y$, $u_{22} = \partial^2 u/\partial y^2$ and likewise v_1, v_2, v_{11}, v_{12}, v_{22}.

At any point of the (x,y)-plane the Jacobian matrix

$$J = \begin{pmatrix} u_1 & u_2 \\ v_1 & v_2 \end{pmatrix}$$

is the linear transformation taking a vector (dx,dy) in the tangent space about the point (x,y) into a vector $(dx,dy)J = (u_1 dx + u_2 dy, v_1 dx + v_2 dy)$ in the tangent space about $(u(x,y), v(x,y))$.

We are interested in the existence of direction pairs which are perpendicular in both tangent spaces. In the tangent space about (x,y), any pair of perpendicular lines can be written either in the form $(1,0)$, $(0,1)$ or the form $(z,1)$, $(-1,z)$ for some real z. We seek a pair perpendicular in both tangent spaces: this means that $(z,1)J$ must be perpendicular to $(-1,z)J$. That is, $(zu_1 + u_2, zv_1 + v_2)$ must be perpendicular to $(-u_1 + zu_2, -v_1 + zv_2)$.

Taking the dot product, we have

$$0 = (zu_1 + u_2)(-u_1 + zu_2) + (zv_1 + v_2)(-v_1 + zv_2)$$
$$= Az^2 + Bz - A,$$ (1)

with

$$A = u_1 u_2 + v_1 v_2 ,$$

$$B = u_2^2 - u_1^2 + v_2^2 - v_1^2.$$ (1a)

Whenever A is not zero the discriminant $B^2 + 4A^2$ of the quadratic (1) is positive, implying that there are two real roots z of the equation. This does not mean there are two pairs of perpendicular lines. The product of the two roots is $-A/A = -1$. Then if one line is $(z,1)$, the other is $(-1/z,1)$, parallel to $(-1,z)$, representing not another solution but rather the other element of the perpendicular pair. Thus there is only the unique biorthogonal pair of which I spoke. The finding is at least as old as Tissot's work of 1881.

When A is zero but B is not zero, there is one root $z=0$ of the equation (1), corresponding to the vector $(0,1)$. This is perpendicular to $(1,0)$ in the (x,y) tangent space, while their transforms $(0,1)J = (u_2,v_2)$, $(1,0)J = (u_1,v_1)$ are perpendicular in the (u,v) tangent space, for the form A is just their dot product.

There remains the case $A=0$, $B=0$. Write the latter constraint in the form $u_1^2 + v_1^2 = u_2^2 + v_2^2 = a^2$, say, and write $u_1 = a \cos \theta_1$, $v_1 = a \sin \theta_1$, $u_2 = -a \sin \theta_2$, $v_2 = a \cos \theta_2$ for θ_1, θ_2 to be determined. Then

$$0 = A = u_1 u_2 + v_1 v_2$$

$$= a^2(-\cos \theta_1 \sin \theta_2 + \sin \theta_1 \cos \theta_2)$$

$$= a^2 \sin (\theta_1 - \theta_2),$$

so $\theta_1 = \theta_2$ or $\theta_1 = \theta_2 + \pi$. Then J is one of the forms

$$\begin{pmatrix} a \cos \theta & a \sin \theta \\ -a \sin \theta & a \cos \theta \end{pmatrix} , \quad \begin{pmatrix} a \cos \theta & a \sin \theta \\ a \sin \theta & -a \cos \theta \end{pmatrix} ,$$

either a similitude or a similitude followed by the reflection in the y-axis. In either case, any pair of lines perpendicular in the (x,y) tangent space is also perpendicular in the (u,v) tangent space.

In general, the conditions A=0 and B=0 each specify curved loci in the (x,y)-plane, and points where A=B=0 are intersections of curves, hence isolated points. Examples can surely be constructed where the constraints A=B=0 are in force along extended loci, but their probability of emerging from actual data is vanishingly small. In the language of the theory of flows, the case of isolated points is "generic." Elsewhere than at such points, the equation

$$A \, dy^2 + B \, dx \, dy - A \, dx^2 = 0 \tag{2}$$

specifies a pair of differential fields whose integral curves exist and are everywhere extensible, by the usual existence theorems. At each point the two values of dy/dx have product -1, and thus are everywhere perpendicular directions. The whole biorthogonal grid is specified by the single differential equation (2) whose coefficients are forms in the empirical Thompson map given by (1a).

It is of interest to examine the behavior of solutions of this system around singularities where A=B=0. At such a point, which without loss of generality we may assume to be (0,0) in the (x,y)-plane, J may be taken as the identity matrix of order 2 by a suitable change of coordinates in the (u,v)-plane. We wish to study the variation of directions dy/dx of our field with displacement from (0,0) in a small neighborhood. Write $(x,y) = (e \cos \theta, e \sin \theta)$, where e is small. Then to first order in e,

$$u_1 = 1 + e \cos \theta \, u_{11} + e \sin \theta \, u_{12},$$
$$u_2 = \quad\ e \cos \theta \, u_{12} + e \sin \theta \, u_{22},$$
$$v_1 = \quad\ e \cos \theta \, v_{11} + e \sin \theta \, v_{12},$$
$$v_2 = 1 + e \cos \theta \, v_{12} + e \sin \theta \, v_{22}.$$

Likewise to first order, substituting in the definitions (1a), we have

$$A = e \, (\cos \theta \, (v_{11} + u_{12}) + \sin \theta \, (v_{12} + u_{22})),$$
$$B = 2 \, e \, (\cos \theta \, (u_{11} - v_{12}) + \sin \theta \, (u_{12} - v_{22})).$$

The coefficients of the forms A and B are all zero only in very unlikely circumstances, for instance, if u and v form a pair of conjugate harmonic functions. For all empirical data sets we may assume that the first-order terms are not all zero near the singularity $(0,0)$ and that they govern the behavior of the solutions there.

Let us write $L = v_1 + u_2$, $M = u_1 - v_2$ and let L_1, M_1 be their partial derivatives with respect to x at $(0,0)$ and L_2, M_2 their partials with respect to y there. Then we have $A = e (L_1 \cos \theta + L_2 \sin \theta) = e \cos \theta (L_1 + L_2 \tan \theta)$, $B = 2 e (M_1 \cos \theta + M_2 \sin \theta) = 2e \cos \theta (M_1 + M_2 \tan \theta)$. When we substitute these values in the equation (2) describing the biorthogonal field near $(0,0)$, and divide through by e cos θ, we arrive at

$$(L_1 + L_2 \tan \theta)z^2 + 2(M_1 + M_2 \tan \theta)z = L_1 + L_2 \tan \theta , \quad (3)$$

where $z = dy/dx$. Note that the variable e has dropped out--to first order, the biorthogonal directions are independent of e for e near zero.

Consider those rays $\theta = \theta*$ out of the singularity $(0,0)$ which lie along one or the other of the perpendicular directions at every point upon them. If the vector $(1,z)$ is on the ray $\theta = \theta*$, then $z = \tan \theta*$. Substituting this value in (3), we have as the equation for the azimuth $\theta*$ of these special rays:

$$(L_1 + L_2 \tan \theta*) \tan^2 \theta* + 2 (M_1 + M_2 \tan \theta*) \tan \theta*$$

$$= L_1 + L_2 \tan \theta*.$$

If L_2 is not equal to zero, this is a cubic equation in tan θ*. (If $L_2 = 0$ we may still interpret it as a cubic with one root tan $\theta* = \infty$, $\theta* = \pm\pi/2$.) The cubic has either one real root or three. In the latter case we may assume all roots distinct--the "generic" case, coincident roots occurring only with probability zero. For each value of tan θ*, of course, there are two directions, exactly opposite each other, which are aligned with the biorthogonal pairs at all their points.

The two possibilities occur as the solution curves which do not pass through the singularity are concave or convex to the singularity. (A simple geometric argument shows they cannot be mixed.) The grids then appear as one or the other of the specimens in Fig. VI-7. These correspond to fields of index -1 and +1, respectively, as described

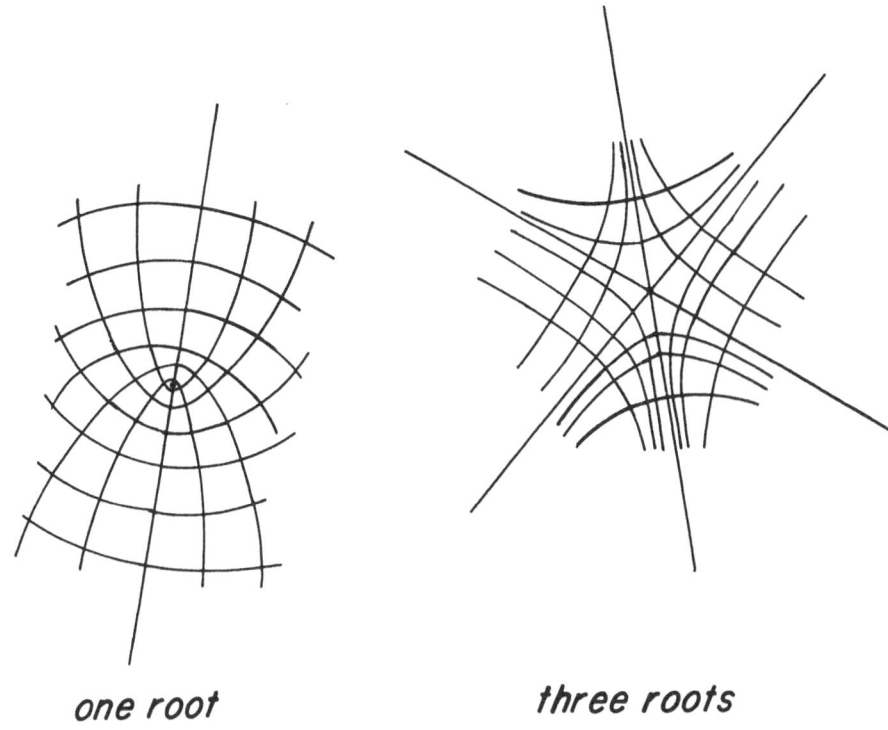

one root   *three roots*

Fig. VI-7.--The two generic singularities for biorthogonal coordinate systems.

in Hopf (1955:30-34). In empirical analyses, either of these occurs
from time to time.

At all points other than these singularities, the appearance of
biorthogonal grids in a small neighborhood is that of a regular network, a so-called "orthogonal parameterization," the distorted image
of regular graph paper, as in Fig. VI-6. Cf. do Carmo (1976), p. 183
(corollary 2), p. 187 (problem 9). Furthermore, all coordinate
curves of a typical grid begin and end only at the boundaries of the
forms. For if they did not, they would have to either end at a
singularity of the system or cycle around a singularity in a spiral
or a loop; and the two generic types of singularities do not allow
this behavior.

It should be noted that there is nothing biologically singular
about the singular points. They are merely loci where our coordinate system breaks down, for purely mathematical reasons.

Technical Note 2. Interpolation from Landmark Locations and Arcs

In computing these biorthogonal curve systems from data it is
necessary to know in advance the one-to-one correspondence (x,y) →
(u,v) between the images. Matters are simplest if one assumes a correspondence between boundaries and extends it to the interiors under
the guidance of any data there. (For an alternate approach, see
Tobler, 1977.) I therefore have adopted the following strategy.
The homology between two images is to be described by discrete pairs
of homologous landmarks, like the corners of the quadrilaterals in
Fig. VI-6, and conic arcs between. (The relation of such a construction to conventional biotheoretical notions of homology is set forth
in Technical Note 4.) Along corresponding arcs the homology is presumed linear. This boundary correspondence, together with any interi-
or point homologies supplied by the data, is imagined mathematically
to distort the inside of either image into pointwise homologous correspondence with the inside of the other. A convenient model for
this correspondence is a complex biharmonic function, with only isolated "sources" or "sinks" of distortion. Selecting one image as the
domain of the coordinate transformation, we may define the u- and
v-coordinates of the mapping function as separate solutions of the
biharmonic equations on that image. The remainder of this Note jus-
tifies this technique and explains it in detail.

The measure of roughness. Algorithms for the "computation" of
an interpolating function usually select, instead, from within a pre-
scribed vector space of functions. That space may have dimension
much higher than the dimension of the data supplied, so that an entire subflat of the function space will be found to fit the data
exactly, and selection of a particular interpolant must proceed via
some ancillary condition. It accords with the biotheoretical founda-
tions of this problem that the computed interpolant should in all
cases be as "smooth" as possible. I shall set forth a measure of
"roughness" instead, which shall be minimized.

Let there be given a pair of shapes C, C' and a biological
homology in the form of a distortion function mapping between them.
A suitable operational interpretation of "smoothness" is as follows.
Consider a small dot square in C and its image in C', as in Fig. VI-
8(a). I suggest a map be considered smooth if the center of the Cartesian square maps into the centroid of the distorted square. As in
Fig. VI-8(b), roughness will be characterized for this little cell by

the squared distance between the image of the centroid and the centroid of the image configuration.

Roughness has a simple analytic approximation. Let the input
square have corners of coordinate $(x_0 \pm h, y_0 \pm h)$ and let the mapping
function $(x,y) \rightarrow (u,v)$ be expressible around (x_0, y_0) as a Taylor
series of second order. Then the four corners of the image configuration are, to second order in h,

$$[u(x_0,y_0) + h(\pm \partial u/\partial x \pm \partial u/\partial y) + h^2/2(\partial^2 u/\partial x^2 + \partial^2 u/\partial y^2 \pm$$

$$2\partial^2 u/\partial x \partial y),$$

$$v(x_0,y_0) + h(\pm \partial v/\partial x \pm \partial v/\partial y) + h^2/2(\partial^2 v/\partial x^2 + \partial^2 v/\partial y^2 \pm$$

$$2\partial^2 v/\partial x \partial y)]$$

of centroid

$$(u(x_0,y_0), v(x_0,y_0)) + 2h^2(\nabla^2 u, \nabla^2 v)$$

where ∇^2 is the Laplacian operator $\partial^2/\partial x^2 + \partial^2/\partial y^2$ and all derivatives
are taken at (x_0,y_0). The value of the roughness is then just
$4h^4((\nabla^2 u)^2 + (\nabla^2 v)^2)$, to fourth order in h. A suitable quantity independent of the scale h of the square is just $(\nabla^2 u)^2 + (\nabla^2 v)^2$, which
can be shown to be invariant under rotation of the coordinate system
for either shape.

In the algorithms that follow, all distortion functions will be
required to have minimum net roughness in the set of all functions
affording the requisite homologies. This is an explicit generalization of certain characterizations of splines. In one dimension, the
familiar cubic spline has the minimum of $\int (y'')^2$ dx subject to constraints at knots and at the boundary (Ahlberg, Nilson, and Walsh,
1969). The integrand here is the square of the numerator of the simple curvature of the function $y(x)$ as a locus in the plane. The
minimum is attained for a piecewise cubic, which perforce has $y^{iv} = 0$
except at the knots. For the two-dimensional, complex analysis, $(y'')^2$
is generalized to $|\nabla^2 f|^2$. It can be shown by use of Green's formula
(cf. Briggs, 1974, or Collatz, 1960:v.5.6) that any function which
minimizes $\int |\nabla^2 f|^2$ over a region satisfies $\nabla^2 \nabla^2 f = 0$ there. This
neatly generalizes the finding that $y^{iv} = 0$ for minimum-curvature
one-dimensional splines. Now $\nabla^2 \nabla^2 f = 0$ is the biharmonic equation,

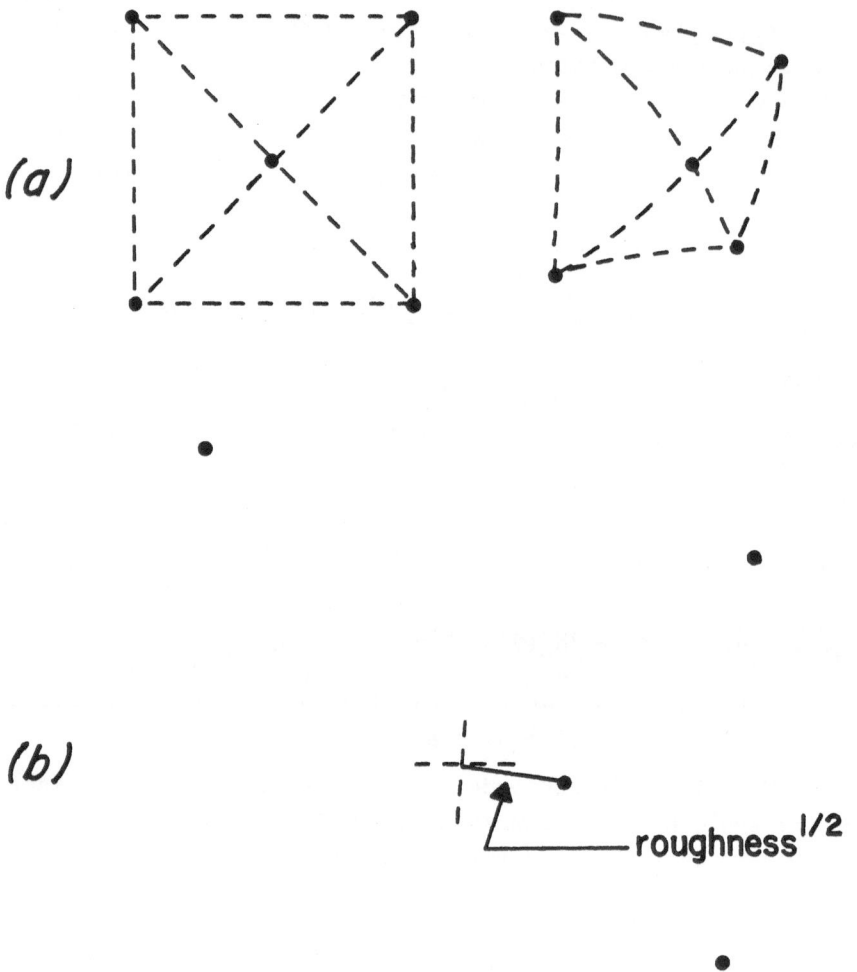

(a)

(b)

roughness$^{1/2}$

Fig. VI-8.--(a) A small square and its image. (b) Enlargement of the right-hand side of (a). Roughness is the squared distance between the image of the centroid and the centroid of the image.

the equation satisfied by thin elastic metal sheets subject to point displacements, just as the spline equation was originally derived to model one-dimensional elastica pinned to a draftsman's board. That my interpolation function satisfies the elastic equation is not a postulate of the model, however, but a deeper consequence of the more intuitive, data-oriented mathematical formulation in terms of smoothness of interpolation.

The vector space and its associated functions. Let C be some simple smooth curve in the plane. Let h be some small spacing, and let (x_0, y_0) be some fixed point. Consider the set S of points (x,y) properly inside C which satisfy the equations $x - x_0 = hn$, $y - y_0 = hm$ for some integers n, m. S is in fact the restriction of a square lattice to the inside of C, as in Fig. VI-9(a). The number of points of S will be denoted by N.

For later use we shall need the set $S_i \subset S$ of "interior points" of S, points all four of whose nearest lattice neighbors lie in S. Denote by N_i the count of points in S_i.

We represent the N-dimensional vector space W over the complex numbers $\mathbf{C}$ by assigning a complex number to each point of S. A basis for this space may be taken as the set of vectors b_s equal to $(1,0)$ at the point s and $(0,0)$ at all other points of S. For any vector v of W, on another picture plane we place the points whose complex co-ordinates are the very numbers v_s assigned to points s of S by the vector v, that is, its components in the basis $\{b_s\}$. These points form a set S', shown in Fig. VI-9(b).

A gnomon G of lattice points may be placed about S such that the set $S \cup G$ includes all lattice points which are within one lattice step of S in any of the eight cardinal directions (north, northeast, east, southeast, south, southwest, west, northwest).

To each vector v in W we associate a smooth function f_v defined throughout C, which has continuous derivatives of first order within C, as follows. Each point of C lies inside some square of side h whose corners are all points of $S \cup G$. Suppose that for every point of $S \cup G$ there is supplied a (complex) value of the function f_v, its derivatives $\partial f_v / \partial x$ and $\partial f_v / \partial y$ in the x- and y-directions, and its mixed derivative $\partial^2 f_v / \partial x \partial y$. Within each square $\{(x,y) \mid x_0 \leq x \leq x_0 + h, y_0 \leq y \leq y_0 + h\}$ the function f_v will be set to the "Coons patch":

$$f_v(x,y) = [u^3 \ u^2 \ u \ 1] \ M \ B \ M^t \ [w^3 \ w^2 \ w \ 1]^t. \tag{1}$$

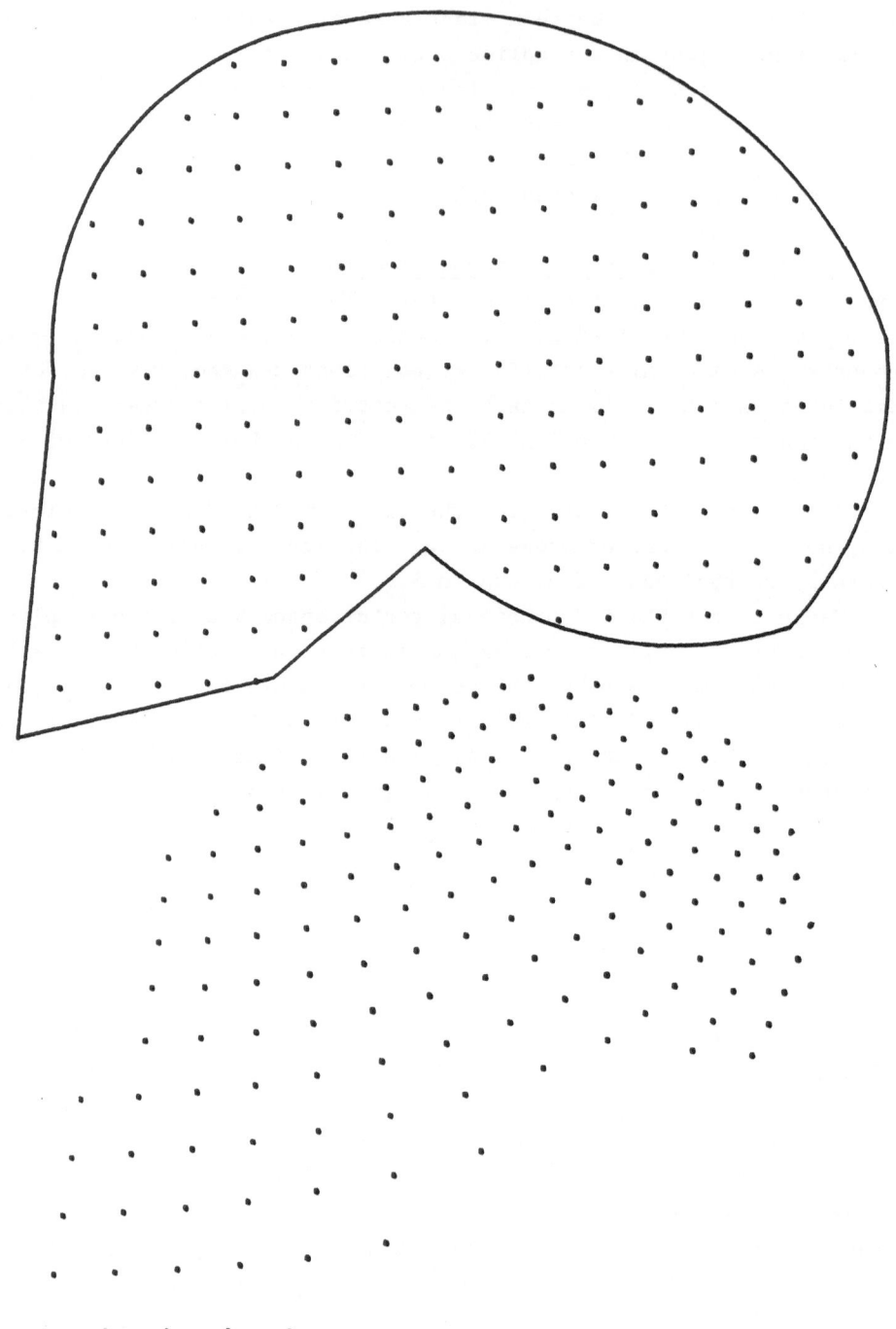

Fig. VI-9.--Conventional diagram of the vector space and one of its vectors. (a) The set S for pre-assigned boundary C, origin (x_0, y_0), and spacing h. (b) The set S' which depicts a particular vector defined "over" S.

Here u = $(x-x_0)/h$, w = $(y-y_0)/h$, M is the matrix $\begin{pmatrix} 2 & -2 & 1 & 1 \\ -3 & 3 & -2 & -1 \\ 0 & 0 & 1 & 0 \\ 1 & 0 & 0 & 0 \end{pmatrix}$,

and B is a matrix of complex numbers selected from those presumed supplied:

$$B = \begin{pmatrix} f_v(P) & f_v(Q) & \partial f_v/\partial y|_P & \partial f_v/\partial y|_Q \\ f_v(R) & f_v(S) & \partial f_v/\partial y|_R & \partial f_v/\partial y|_S \\ \partial f_v/\partial x|_P & \partial f_v/\partial x|_Q & \partial^2 f_v/\partial x \partial y|_P & \partial^2 f_v/\partial x \partial y|_Q \\ \partial f_v/\partial x|_R & \partial f_v/\partial x|_S & \partial^2 f_v/\partial x \partial y|_R & \partial^2 f_v/\partial x \partial y|_S \end{pmatrix}$$

where P is the point (x_0, y_0), Q = (x_0, y_0+h), R = (x_0+h, y_0), and S = (x_0+h, y_0+h). This function f_v is the unique bicubic polynomial in the Cartesian coordinates which satisfies sixteen corner conditions, namely, assigned values of f_v, $\partial f_v/\partial x$, $\partial f_v/\partial y$, and $\partial^2 f_v/\partial x \partial y$ at each of the four corners of each little square. (Cf. Forrest, 1972, and Rogers and Adams, 1969: ch. 6.) It can be shown that this function is C^1 at all lattice points and on all vertical or horizontal segments through lattice points. Inside each square, of course, it is C^∞ by virtue of its polynomial form.

The values of f_v on S are available--they are the components v_s of v, the points of S'. We estimate values of f_v upon the gnomon G as well, by extrapolation along the eight cardinal directions. With values for f_v now assigned at every point of S $\cup$ G, values of $\partial f/\partial x$, $\partial f/\partial y$, and $\partial^2 f/\partial x \partial y$ are assigned by conventional formula (cf. Abramowitz and Stegun, 1964:879,914).

This algorithm may be compared with that of Akima (1974), which is based on nonlinear estimators. Neither my technique nor his should be confused with ordinary bicubic spline interpolators (cf. Schumaker, 1976), whose values are determined globally by all the values f_v together with certain boundary conditions. These bicubics themselves minimize certain integrals in $|\nabla^2 f|^2$; in the algorithm I present it is the vector, not the function f, which bears optimality properties.

Values of f_v and its derivatives now having been set throughout

S U G, the values of f_v anywhere in C may be computed square by square according to the interpolation formula (1) preceding. Since f_v maps each point of S into the corresponding point of S'--the point v_s carried by the sth component of v--it may be considered a C^1 interpolant of the mapping S $\rightarrow$ S' as a function defined throughout C.

These functions f_v are themselves clearly a vector space over C isomorphic to W. For each s, the formulas for the derivatives are linear on W, as is the formula for interpolation within lattice squares. Then for fixed (x,y) ε C, lattice point or no, the value $f_v(x,y)$ is a linear functional on W. We will use this fact presently.

It is useful to consider an alternate basis for the vector space W, one naturally associated with the roughness measure. Let s be any point of S_i. Then the sth element of the alternate basis is the vector

$$r_s = b_s - \tfrac{1}{4} \overset{\displaystyle}{} b_{s'}$$

s' neighbors of s
north, east, south, west

Let M_s be the vector of coefficients of r_s in the ordering of S; in general it looks like

$$(0, \ldots, 0, -\tfrac{1}{4}, 0, \ldots, 0, -\tfrac{1}{4}, 1, -\tfrac{1}{4}, 0, \ldots, 0 -\tfrac{1}{4}, 0, \ldots, 0). \qquad (2)$$

A well-known approximation in numerical analysis states that
$$\nabla^2 f_{(x_0,y_0)} \sim h^2 f(x_0,y_0) - \tfrac{1}{4}(f(x_0+h,y_0)+f(x_0-h,y_0)+f(x_0,y_0+h)+f(x_0,y_0-h)).$$

In the current notation, this means that the roughness of f_v at s is approximately $|r_s v^t|^2$, where $|z|$ is the Gaussian norm $\sqrt{z\bar{z}}$. I shall hereinafter identify roughness with this value.

For points of S not in S_i, the formula for roughness at s cannot be that given above. The usual approximation is instead the Shortley-Weller (Proskurowski and Widlund, 1976:442). If s is not in S_i, it has one or two neighbors outside C. The segments from s to those neighbors cut the boundary of S as shown in Fig. VI-10. At such cuts there will eventually be assigned scalar values b_1 and b_2, "boundary values," derived from the one-to-one correspondence between boundaries. On either arm of the cross through s, through the three values unevenly spaced there passes a unique quadratic polynomial, which, extrapolated, yields a value to assign to the missing neighbor

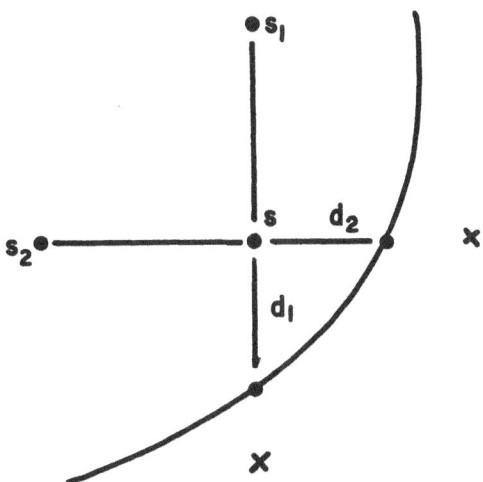

Fig. VI-10.--Geometry of the roughness computation near the boundary of S.

outside S. From the reconstructed full-size cross the roughness may be computed as usual; it is a Hermitian form in the components and the scalars. For the case illustrated in Fig. VI-10, the approximate roughness is the square of the norm of

$$-2v_{s_1}/1+d_1 - 2v_{s_2}/1+d_2 + (2/d_1 + 2/d_2)v_s$$

$$- 2b_1/d_1(1+d_1) - 2b_2/d_2(1+d_2), \tag{3}$$

where d_1, d_2 are the distances to boundary shown. Should only one neighbor of s be outside S, the formula is modified by setting the other d to unity and replacing the boundary value (a scalar) by the appropriate v_s (a component).

The homogeneous part of the formula (3) supplies a vector M_s for the points s of $S-S_i$, just as (2) supplies M_s for $s \epsilon S_i$. Construct the matrix M, which is NxN, by $M = (M_1{}^t|M_2{}^t| \ldots |M_N{}^t)$. Most monographs on the numerical solution of partial differential equations prove that this matrix is nonsingular. (See, for instance, Collatz, 1960: I.5.5, theorem 2 and p. 348.)

Interpolation from boundary values. Let B be the vector in W, of components indexed by s in S, which has entries 0 for s in S_i, and

the appropriate nonhomogeneous term of the formula like (3) for s
not in S_i. Consider the system of equations $v_0 M = B$, where M is the
matrix defined just above. Since M is nonsingular, we can solve this
system of equations for v_0: $v_0 = BM^{-1}$. By the manner in which M and
B were constructed, the vector v_0 can be characterized as the unique
vector which, in conjunction with the boundary values which entered
into the components of B, has approximate roughness exactly zero at
every point of S.

Suppose, now, that the boundary values entering into the components of B are assigned as boundary <u>points</u>, themselves complex numbers, of a second shape C'. The assigned values are to be the exact
homologues, under some conventional interpolation, of the boundary
points of C which served as surrogate neighbors in the Shortley-Weller
formula (3). Then the set of components of v_0, plotted as points
making up a set S', will depict a distortion of S which interpolates
into the interior of C' with roughness exactly zero, and the function
f_{v_0} will be just the distortion function needed in the D'Arcy Thompson formulation for this mesh.

The entire approximating procedure will be well-defined once
the boundary values are assigned. In my implementation, the boundaries of shapes are constrained to be segments or arcs of conic sections among a small number of landmarks. Between corresponding
segments the homology between boundaries is linear in distance; between corresponding arcs it is linear in arc-length.

<u>Interpolation from boundary values and interior points</u>. The
boundary-driven interpolation function f_{v_0} has roughness zero at
every point of S. This function maps those boundary points on grid
lines very nearly onto their homologues; it maps the boundary <u>land-
marks</u> onto their homologues to the extent that they are represented
by the nearby boundary points on grid lines. But the function f_{v_0}
takes no cognizance at all of interior points--as indeed it cannot,
for its vector v_0 is the solution of an exactly determined linear
system, with no degrees of freedom to spare.

Insisting that a set of interior points of C be mapped exactly
onto homologues in C' will require a modification of f_{v_0}. To this
end, consider the computation of v_0 not as the solution of a linear
system but as the minimization of the total roughness $|vM - B|^2$, where
$|\cdot|$ is now the Hermitian vector norm, $|(v_1, \ldots, v_N)| = \sqrt{\Sigma v_i \bar{v}_i}$. The

unconstrained minimum of this form is zero, attained at the value v_0 just computed.

To homologies on interior pairs correspond constraints on the vector v. Each constraint will be of the form $f_v(x_1{}^i, y_1{}^i) = (x_2{}^i, y_2{}^i)$. By the construction of the function f_v, its value at $(x_1{}^i, y_1{}^i)$ is a linear combination in the components of v, the values of f_v at the grid points around $(x_1{}^i, y_1{}^i)$. The coefficients of this linear combination are functions of the relative location of $(x_1{}^i, y_1{}^i)$ in the grid square surrounding it and the proximity of those grid points to G.

Let us write $f_v = f_{v_0} + f_{v_1} = f_{v_0} + v_1$, where f_{v_0} is the interpolant corresponding to zero roughness, boundary determination only, and f_{v_1} is an adjustment for interior points. We must have

$$f_{v_1}(x_1{}^i, y_1{}^i) = (x_2{}^i, y_2{}^i) - f_{v_0}(x_1{}^i, y_1{}^i)$$

for every pair of points constrained to correspond. Since f_{v_0} fits the distortion very accurately on the boundary of C, that, at the intersections of the boundary of C with grid lines, we also must have f_{v_1} identically zero there. We wish to compute the vector v_1 subject to these constraints for which the roughness of f_{v_1}, which is the same as the roughness of f_v, is minimal.

For the estimation of v_1, the appropriate vector B is identically zero, and the roughness of f_{v_1}, summed over all the points of S, is just equal to $(v_1 M)(\overline{v}_1 M)^t$. We wish the vector for which this quantity is minimal, subject to a set of constraints regarding interior pairs $(x_1{}^i, y_1{}^i)$, $(x_2{}^i, y_2{}^i)$. Let the constraints be assembled in the single matrix form $v_1 Q = R$, where Q is the sandwich of coefficient column vectors expressing each $f_v(x_1{}^i, y_1{}^i)$ as a weighted sum of the components of v, and R is a vector with ith entry $(x_2{}^i, y_2{}^i) - f_{v_0}(x_1{}^i, y_1{}^i)$.

Let us change to the alternate basis by writing $v_2 = v_1 M$. The problem is now to calculate v_2 such that $v_2 \overline{v}_2{}^t$ is minimal subject to $v_2 M^{-1} Q = R$. This is a standard problem in quadratic forms (cf. Rao, 1973: sec. 1f.1); the solution is at the orthogonal projection of the origin onto the flat $v_2 M^{-1} Q = R$, which is

$$v_2 = R((M^{-1}Q)^t (M^{-1}Q))^{-1} (M^{-1}Q)^t .$$

The vector v_2 supplies an imputed roughness at every point of s; the actual complex loci for v_1 are gotten by reversing the change-of-basis-- $v_1 = v_2 M^{-1}$. For this v_1, the function $f_{v_0+v_1}$ maps C onto C', approximately preserves the boundary homology and the homology of scattered interior pairs, and minimizes roughness over the set of all v whose f_v can be so described. It is the smoothest distortion function possible for applications of the Thompson method to data which include interior correspondences.

Difficult configurations for multiple inside points, for instance when two points exchange positions from shape I to shape II, lead to a homology map which does not have a single-valued inverse. This can be checked by a glance at the output grid, which will show shock waves and catastrophes. Automatic interpolation by optimal linear smoothing simply does not apply in cases of distortion so extreme.

Note on computation. The fast computation of the product of M^{-1} by arbitrary vectors is carried forth using published code which implements the so-called capacitance matrix method. Proskurowski and Widlund (1975, 1976) describe and list a computer program which solves this equation vM=T for arbitrary real vectors T and a quite general class of object geometries (sets S) by an easy series of calls from a Fortran driver. By executing the computations twice, once with the real part of the boundary vectors and interior homologues, again with the imaginary parts, the complex problem is exactly solved; for the whole estimation of v is separable, roughness and all, into its real and imaginary components.

The principal axes at any point are constructed from this mapping function and its derivatives there according to the formulas of the preceding Note.

Technical Note 3. Construction of Integral Curves

The biorthogonal grids in this monograph were all produced by a computer program running under the MTS operating system on the Amdahl 470V/6 computer at the University of Michigan. The program's acquisition of outline pairs does not concern us here. Once a Cartesian interpolation is completed according to the system of the preceding Note, the user may wish to summarize it by a suitable sample of curves from its biorthogonal grid. Sitting at a cathode ray tube, he indicates an arbitrary sequence of points inside one form; through these points the program computes integral curves of the tensor field of local principal strains that is implicit in the interpolation function. The curves are drawn out superimposed over the boundary drawings, with the appropriate dilatation gradients sampled along them. The user may delete curves and replace them with others to make a more balanced composition, then send a final image to a Cal-comp pen plotter for "hard copy." All in all, a very satisfying procedure, "hands-on" and fun to watch.

It remains to set forth the algorithm for constructing these curves. Points mentioned in the paragraphs following are displayed, for a typical computation, in Fig. VI-11.

Let m_0 be a point inside image I, and v a unit vector along one of the four principal directions through m_0. The point m_0 is con-sidered a "node" of the curve if the algorithm starts there; the out-put is a new node m_4 along with a "midnode" m_4' making for gentler curves in the drawing.

Let the line L be constructed perpendicular to the vector v through the point $n = m_0 + d \cdot v$ displaced from m_0 along the assigned direction at a default spacing d. If the integral curve through m_0 aligned with v is approximately a circle, which we assume only tem-porarily, then near the point n there should be a point m_1 on L such that the principal direction v_1 at m_1 is at the same inclination to

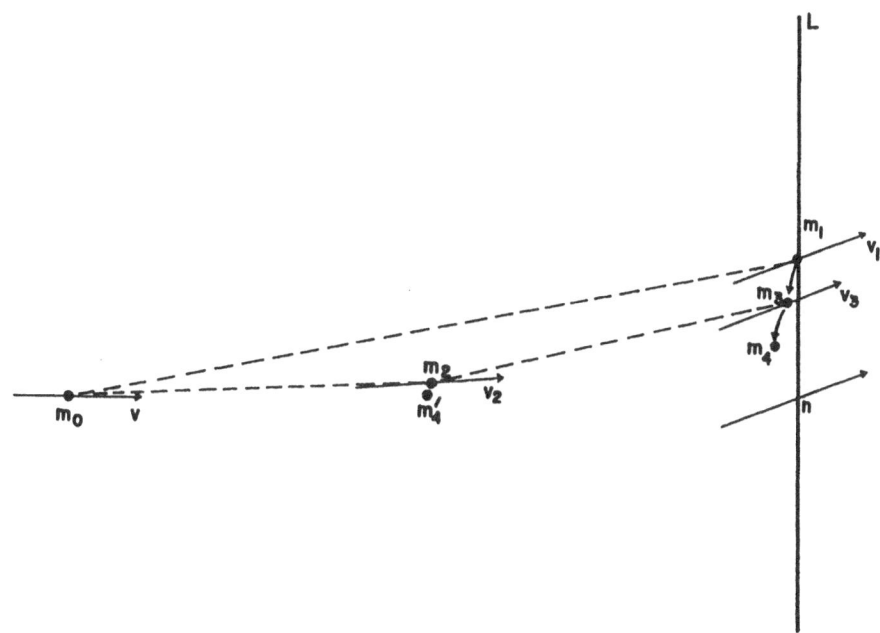

Fig. VI-11.--One step in the integration of a line of growth.

the chord $m_0 m_1$ as is the vector v. This point m_1 is estimated by iteration.

The estimation is repeated with d replaced by d/2, to compute a point m_2 and a vector v_2 approximately half as far out from m_0 along v. We then replace m_0 with m_2 and v by v_2, leaving the spacing at d/2, to compute a point m_3 near m_1 "twice as precisely."

This dependence of the final point, m_1 or m_3, on the scale of the computation, d or d/2, is now extrapolated to d = 0. There result the following two points, which are presumed to be on the integral curve through m_0:

$$m_4 = m_3 + (m_3 - m_1), \text{ the next node; and}$$

$$m_4' = m_2 + .25(m_3 - m_1), \text{ the "midnode."}$$

The images of m_4 and m_4' under the mapping function are likewise assigned to an image integral curve through the homologue of m_0.

The curve fragment $m_0 - m_4' - m_4$ is extended further by replacing m_0 with m_4, replacing v with the exact principal exis computed at the new m_0, and executing the algorithm again.

Should the estimation of any of m_1, m_2, or m_3 fail to converge, or should principal directions at any of the m_i deviate far from the pre-assigned tangent v, the spacing d is cut and the procedure is begun over with the same m_0 and v. If there is still no convergence, the curve is deemed simply to stop at m_0, probably owing to proximity to a singularity of the coordinates. But curves do not necessarily cease at singularities. A path aligned accurately with the special rays of azimuth tan $\theta*$ discussed in Technical Note 1 may jump over, if the spacing d is suitable, and proceed, undeviated, on the other side. Certain curves of Figs. VII-3 ff. do just this. Except for cessation at singularities, the integration proceeds until the integral curve crosses the boundary of the images.

Technical Note 4. On Homologous Points

The algorithm preceding depends upon a notion of homology expressed wholly in terms of mathematical points. This may appear to be quite different from the version current in conceptualizations of biology, but is in fact concordant, indeed, is a superordinate formulation.

In presenting the conventional view I excerpt the work of Jardine (1967, 1969). In his formalism, homology is a matter of pairs

of "parts, organs, or structures" and of relationships among them
mostly spatial--"anterior to," "distal to," "ventral to," "inside of,"
and their opposites--between parts. A homology between two organisms
is defined to be a maximally inclusive scheme of pairs that manifest
the same positional relations among themselves in both organisms.
Such a homology is "computed" by a combinatoric search over sets of
pairings for more and more extensive schemes. The computed homology
is explicitly dependent upon both the two lists of parts and the list
of spatial relations whose correspondence the researcher chooses to
insist on. The list of parts, in particular, may include fused structures, or extended curves, like canals, which pass through structures,
and embody thereby a more or less subtle notion of the relatedness of
the organisms that are being studied.

Jardine does not scrutinize at all closely the geometric assumptions and structures underlying his method. The "parts" he analyzes
are chunks of biological tissues, but insofar as he is analyzing spatial relations certain topological conditions of real Euclidean space
are quite relevant. He acknowledges a point made by Sneath, that a
part is detectable only by some sort of qualitative discontinuity at
its boundary, after the manner of a bone suture. In fact, the spatial properties to which Jardine makes reference are all properties
of the boundaries alone; the qualities of the interiors are not used
at all (for in fact they are entirely bone in his examples). In his
diagrams, these boundaries are all simple closed curves in the plane,
curves containing a few special points where sutures intersect,
canals cross in or out, and the like. There are also special boundary lines of possible fusion or loss which might be missing in one
form or the other.

In Jardine's procedure there is considerable imprecision of measurement. He is dealing with regions, but the spatial relations he in-
vokes--"anterior to" and the others--are relations on pairs of points.
The points of a region may not be in constant spatial relation to
points of another region. Jardine himself notes (1969:334) that
there is an ambiguity of scoring created thereby. He adjusts to this
possibility by making spatial relation into a ternary variable, add-
ing "undecided" to the classes "yes" and "no." But this device does
not extricate us at all from the conceptual difficulty. In the arrangement $\begin{array}{cc} A & B \end{array}$, the "ambiguous" one, there is in fact
no geometric ambiguity of anteriority. The points A, B, C, D are in

perfectly well-defined order (ABCD).   (BADC) is the "opposite" ambiguity.   The orders (ACBD), (BDAC) Jardine would call unambiguous;. there
are (ABDC) and (BACD) as well, "ambiguities" of a different style.
But none of these are ambiguous; all are legitimate orders of points
of equivalent intrinsic interest.   The point basis of spatial relations is implicit even in the analysis of a single region.   It is not
really the scalar spatial relations, like anteriority, which Jardine
needs at all.   If we identify special points, "landmarks," on the
boundary of a "part," then a correspondence of points will be a
satisfactory homology if around each boundary the corresponding landmarks are encountered in the selfsame order in both images

Jardine refers in passing to the possibility of an "appropriate" spatial distortion, a "grid transformation" (= mapping) after the fashion of D'Arcy Thompson, ancillary to and prior to the
evaluation of spatial relations.   Such a transformation, while altering none of the order relations I have been discussing, would adjust
the local axes by which anteriority, etc., are measured, and allow
for gross topographic changes among the putative homologues.   Now, as
it happens, the class of all possible transformations is identical
with the class of all one-to-one differentiable mappings of one organism onto the other, and this is a very large mathematical class
indeed.   We may map any finite set of points onto any other set, in
any permutation, by such a distortion.   To the set of standard boundary points we might add some other assortment of special points on
the inside of regions, such as presumed ossificatory centers, and our
ᴸreedom to map is still the same.

Given such a transform, to any geometric region (part) there
corresponds a homologous region.   We need to compute that correspondence which comes closest to matching parts with those whole parts
that are predetermined--which has the best matching of boundary point
orders.   The resulting mapping could then be adjusted (still maintaining the correspondence of boundary points) so that the actual
boundary curves matched in extenso.   If such a transform could be
arranged, there would be no need for spatial relation measurement
or any other verification: the homology would be by direct reading
of the mapping.   In principle, a computer program could try all reasonable (i.e. order-preserving) permutations for the pairing of landmarks and choose the one with the least global strain appropriately
scaled.   In practice this would be grossly expensive, of course.   But
I submit that the Jardine algorithm is a very rough equivalent.   The

regions are substitutes for the average grid-coordinates of their
boundaries. The combinatoric of the Jardine scheme is phrased
coarsely, in terms of regions; in fact, it ought to be of a larger
point-set upon the spectrum of possible one-to-one differentiable
mappings. The "spatial relations" are simply differences of coordinate in the appropriate grid coordinate system. This formalism
settles not only the ambiguities arising from overlap of parts but
also the problem of comparing parts arbitrarily far apart. They
simply have corresponding coordinates in the two grids, and it does
not matter in which direction "anterior" is.

The Jardine homology theory, then, based in spatial relationships,
is a transformation grid crudely expressed. Though it is computed in
terms of regions, it is equivalent to a formulation based on points.
Such points, whose existence Jardine does not acknowledge, are exactly
the homologous landmarks of my technique, distinctive loci which correspond and which indicate a correspondence of all the anonymous
points in between them on curves and in interiors. Jardine's homology is just a special case of the shape changes with which part two
of this essay is concerned.

CHAPTER SEVEN.   EXAMPLES OF BIORTHOGONAL ANALYSIS

A.   Comparison of Square and Biorthogonal Grids:
     Thompson's Diodon Figure
     One of Thompson's original examples is famous beyond all others:
the comparison of Diodon with Orthagoriscus (= Mola) in Fig. V-2.
Thompson's description of his grid pair has already been excerpted on
page 70.   For reanalysis, I have represented the two outlines by
seven point pairs and circular arcs connecting them.   The points include the mouths and the axial ends of cranial and caudal margins of
the dorsal and anal fins, together with points which correspond, according to Thompson's drawings, on each of the arcs from mouth to
fins.   The circular arcs pass through ordinary points approximately
midway between consecutive landmark pairs in Thompson's figure.   Of
course, the fins, being outside the resulting boundaries, are omitted
from the computed representation.
     The smooth interpolation of homologies inside the boundaries,
shown by the meshes in Fig. VII-1(a) closely resembles Thompson's
work.   But a biorthogonal summary of this distortion, Fig. VII-1(b),
is strikingly different in structure.   It is clear at a glance that
the "true" lines-of-growth are situated exactly in opposite orientation from the way Thompson set them: they are very square in Mola
but converge in Diodon.   The back of Mola is pushed out in fields
concentric from the tail, not from the head.   The shear which Thompson attributes to "friction or restraint" is in reality only an
artifact of his faulty choice of coordinates.   In certain other features, also, the drawings disagree.   Diodon has a deeper throat than
Mola.   The most ventral lines-of-growth indicate this by converging
toward the head in Mola but not in Diodon.   Thompson did not recognize this asymmetry at all in his grid, though it is clear in his
data.   Furthermore, if we extract approximate local dilatations by
ratio of corresponding lengths, Mola's over Diodon's, in the generally vertical dimension all along the midline, we observe a growth-
gradient as pretty and monotonic as you please, from 1.2 at the anterior end, through 1.6 around the pectoral fin, to 6 near the caudal
fin.   Thompson makes two errors relative to this feature.   He draws
his axes in sudden divergence away from the midline behind the pectoral; and at the caudal fin the dilatation measurable in his axes
is only 4.0.
     Thompson viewed the curving of the caudal margin of Mola as a

major shape difference which propagated inward to bend all the vertical curves 0 - 6 of his grid into "concentric circles." In the
rigorously smooth interpolation here, that bend is present, but its
implications are wholly different. It is possible that the greater
bend in Thompson's drawing derives from the positions of the midline
structures--gill slit, pectoral fin. I have located the centers of

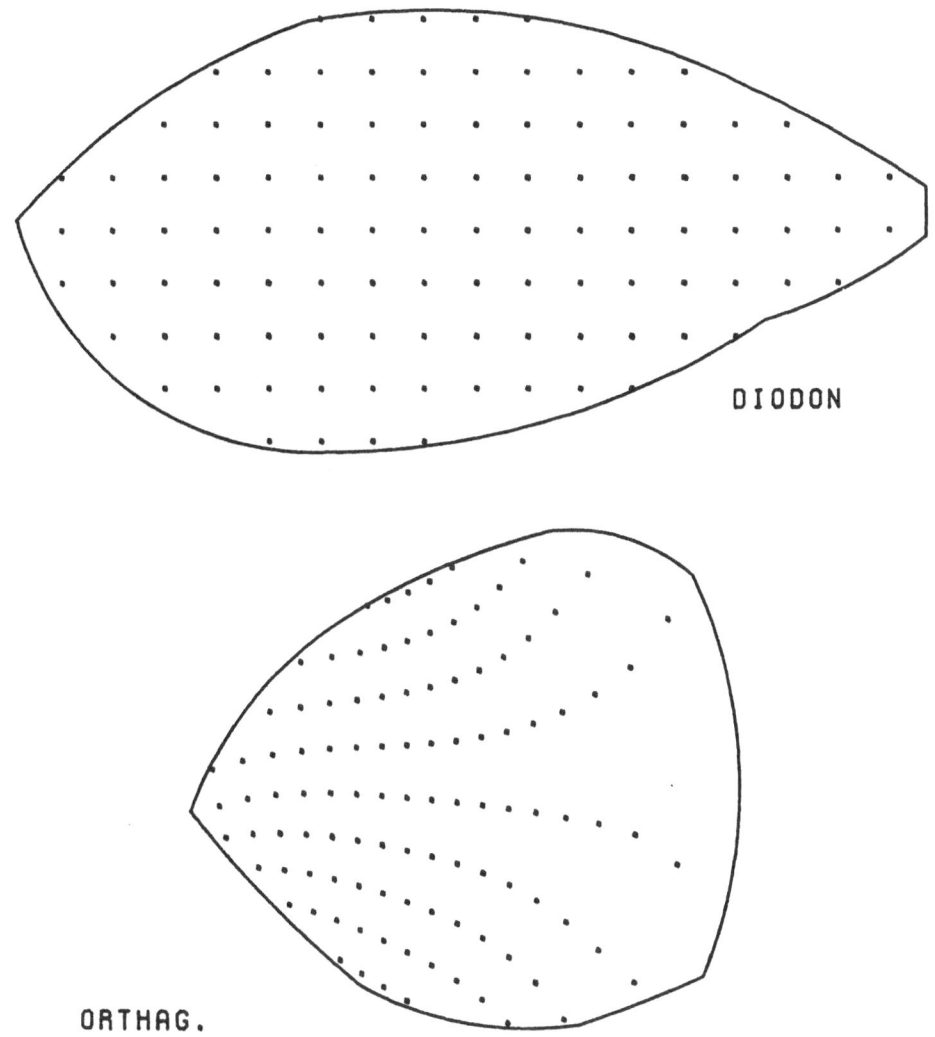

Fig. VII-1(a).--Cartesian grid for the transformation of <u>Diodon</u>
into <u>Mola</u>, making use of boundary homologies only.

the cranial margins of these structures at the "extra" dots, those not conformable with the smooth strings of mesh lines, in Fig. VII-1(c). The lower mesh there results from constraining Fig. VII-1(a) to accord with these two interior homologies, by the algorithm of Technical Note 2. An extract from the corresponding biorthogonal grid, Fig. VII-1(d), indicates that the adjustment has virtually no

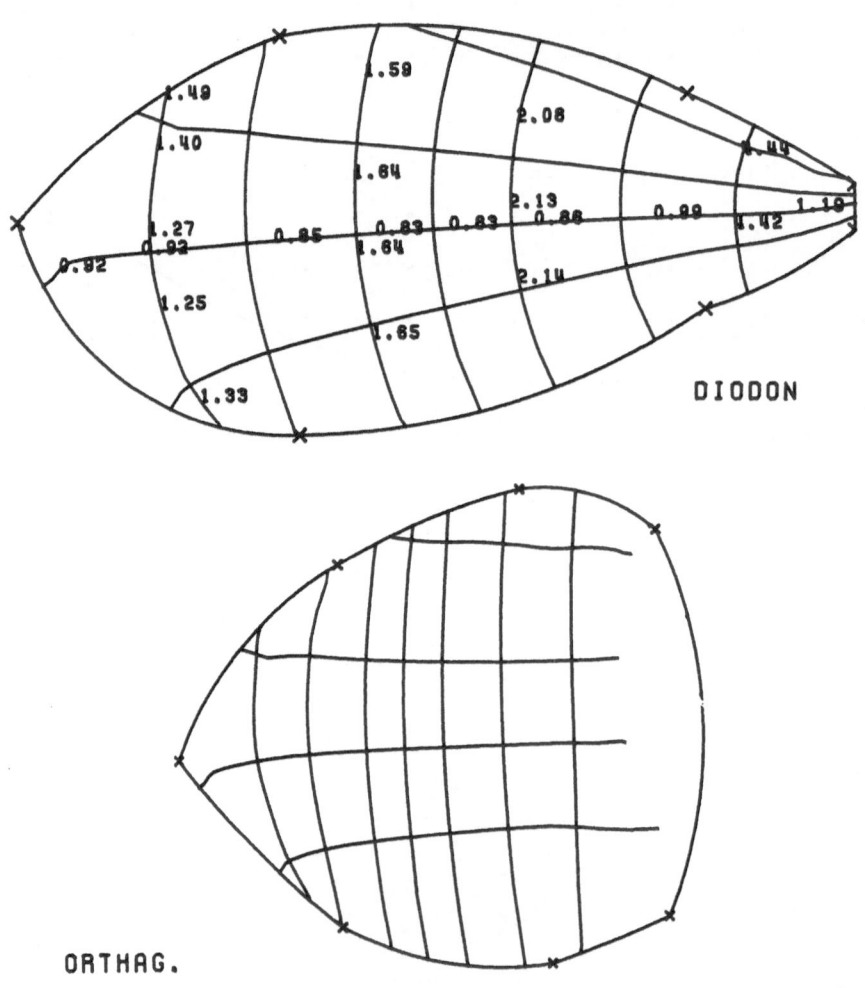

Fig. VII-1(b).--Biorthogonal grids for the same transformation.

effect on the caudal half of the drawing, where the impression of a
morphogenetic field about the tail is preserved. We have merely generated a singularity near the gill where the dorsoventral dilatation,
about 1.2, becomes equal to the craniocaudal dilatation now enhanced
by the "pull" of the landmarks backwards in <u>Mola</u>. Of the two analyses, Thompson's and mine, that by empirical lines of growth clearly

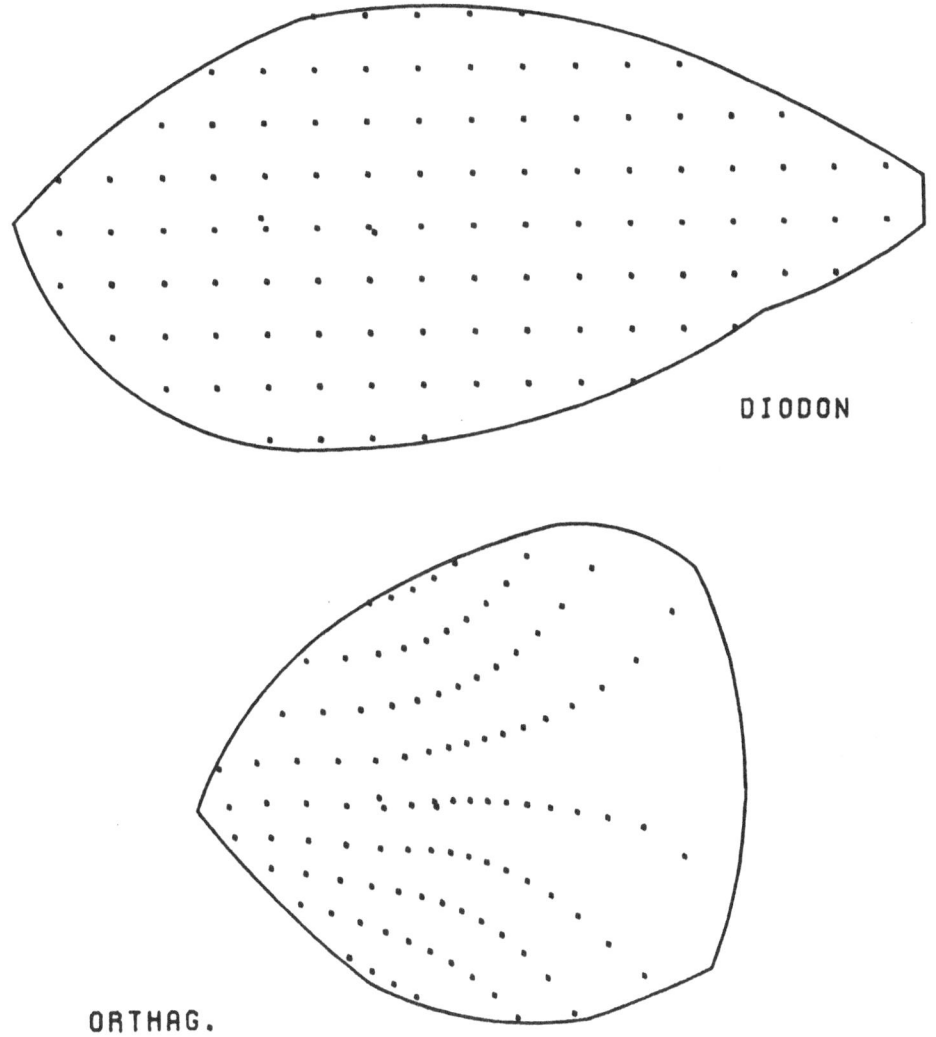

Fig. VII-1(c).--Cartesian grid for the same transformation, cor-
recting for two interior homologies: gill slit and pectoral fin.

supports the more sensitive and reliable shape comparisons.

B. Phylogeny and Ontogeny of Primate Crania
        In this subchapter I wield the biorthogonal method to signify
very suggestively a point of human phylogeny.

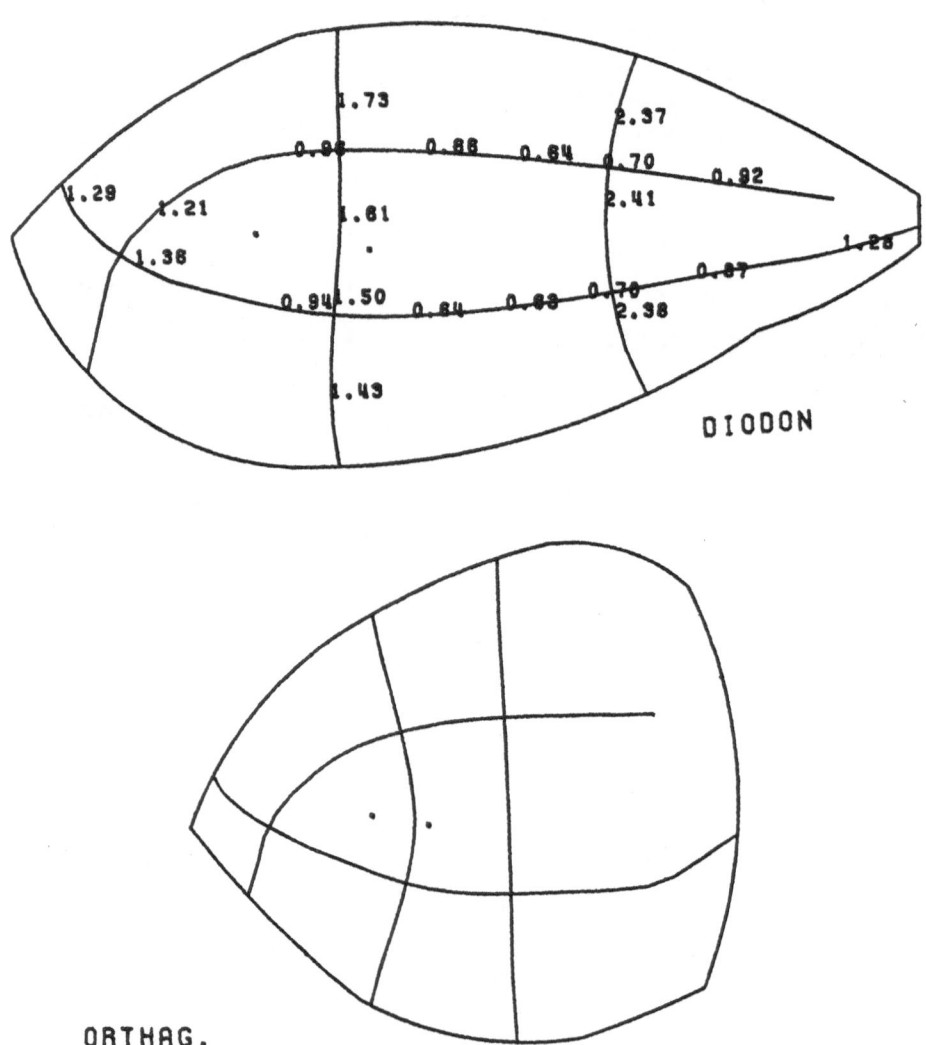

DIODON

ORTHAG.

    Fig. VII-1(d).--Sketch of the biorthogonal grids for the same
transformation.

1. Functional craniology and craniometrics

The higher primates differ anatomically among themselves, we humans from apes in particular, almost wholly in relative proportion of parts and structures rather than their presence or absence. Yet peculiarities of human function--bipedalism, rational intelligence, and the like--seem to drive the adjustment of form uncomfortably beyond the limits of easy assimilation. The ordinary prime movers of evolution, allometry and the adjustment of relative rates of growth (cf. Gould, 1977: chs. 7-9; Rensch, 1960: ch. 6.B.iii), are surely sufficient for the crucial changes toward humanness of form; but it is not certain to what extent the known fossils resemble our actual ancestors, nor how many independent complexes of trends there might be.

In the large anthropological/primatological literature on this subject (for reviews, see, for instance, Starck, 1974, or Schultz, 1950), the head is singled out for special study as one of the regions of the body (another is the lower limb) which show the greatest morphological change over recent evolutionary time. Following functional craniology, we assume that this all is recorded in the bones. It is the basic axiom of functional craniology, in fact, that the bony cranium is "plastic" over ontogenetic and evolutionary time, that the cranium as a whole expresses the coordinated alterations of all its attachments--teeth, muscles, brain. I have not the space here to review this assertion in detail. Good surveys of the subject, demonstrating a variety of styles of experimental and comparative evidence, include Klatt (1949), Scott (1963), Biegert (1963), and Moss (1973 and references therein).

These alterations over the course of primate phylogeny may be summarized under the following points, drawn from Biegert (1963) and Hofer (1965):

(a) expansion of the cortex of the brain, which "rolls over" the cranial base, vaulting front and rear--an effect negatively allometric with body size;

(b) expansion of the masticatory apparatus, with associated remodelling of the superstructures of the bony vault--an effect positively allometric with body size;

(c) development of molar grinding, which requires the mandibular joint to lie well above the dental arch, and positions the jaws subcerebrally, or induces an orbital torus, for the distribution of stresses;

(d) rotation of the eyes forward, to facilitate stereoscopic vision;

(e) reduction of the snout (but its replacement by a muzzle in the larger forms);

(f) expansion of organs under the mandible, which tilts and translates the mandible and the structures which articulate with it;

(g) assemblage of the whole into a smooth solid form, structuring the joins between parts by sinuses and other buttressing.

Numerical summary of this diversely generated form change--measurement of the growing, evolving head--is a most intricate metric task, one for which conventional cephalometrics, designed for static comparisons of diverse adult human populations, is unsuited. On hominization, in particular, the literature includes competing single-factor theories which are not brought to bear upon each other at all. The various summary hypotheses include "fetalization" (retardation of ontogenetic timing), an argument reviewed in Gould (1977: ch. 10), and cephalization (increase in relative size of the head's neural contents), a matter discussed profoundly in the work of Weidenreich (1941) and Hofer (1965 and references therein). Each argument is based on a host of simple angles and ratios which are either monotonic with the general grade of hominization, which is fine, or else contradict it, in which case they need to be explained away. Each index has its own particular inappropriateness to such a general explanation, as Schultz (1955) exemplifies for measures of position of the occipital condyles, Biegert (1963) for prognathism.

Central to all of these scalar explanations is the examination of the cranial base, its dimensions and angles. A good introduction to the anatomical facts bearing on this subject is Scott (1958). Along the midline of the head, which is the only section I shall deal with here, the cranial base comprises several structures. Rearmost is the basioccipital bone, which articulates with the atlas vertebra of the spine; basion is the point on the midline at this articulation, the point farthest forward on the foramen magnum. The upper surface of this bone is named clivus. Between the basioccipital bone and the basisphenoid is a cartilaginous segment, the sphenooccipital synchondrosis, which closes (fuses into bone) at about age twenty in humans. Further along the top of the basisphenoid is a little pocket, the sella turcica, in which rests the pituitary gland. Anterior to that is another synchondrosis, the midsphenoidal, which fuses prenatally in the apes and man but stays patent to adulthood in the monkeys and prosimians; and forward of that, past the spheno-

ethmoidal synchondrosis, is the cribriform plate of the ethmoid bone,
which articulates to the frontal bone near <u>nasion</u>, the external suture of frontal and nasal bones. This series of structures functions
loosely as a boundary between the neurocranium and the splanchno-
cranium, between that part of the cranium responsive mostly to brain
development and that part expressing mastication and other functions.
The base is also responsive to the needs of the body expressed at the
pharyngeal space, e.g. breathing (Baer and Nanda, 1976).

In this series of structures there is indubitably a <u>kyphosis</u>,
a bending, during both ontogeny and phylogeny. Its precise quantification is uncertain, as the lines by which it might be measured
can be laid tangent at two places to various bony surfaces or instead
passed through various landmarks, and different computations ensue.
Ontogenetically the bend is traceable to both of the sphenoidal sutures, which are patent for different periods and which appear to
change the sense of their flexure from fetal to postnatal growth.

Whatever the exact locus of the kyphosis, its main effects over
the course of hominization are clear: it permits expansion of the
outer circumference of the brain while the cranial base remains relatively stable in size, the dependent face shrinks, and the foramen
magnum rolls forward, all this assuming no specific compensations.
But the establishment along these lines of a geometric model based
upon comparative data is unexpectedly difficult. First, the relative
position of sella with respect to basion and nasion, or other indi-
cators of the two polarities of this extended structure, is not expressed purely in that angle, but also in the ratio of sella's distances from the two ends. Parts of the cranial base lengthen and,
over evolution, shorten all the while they are rotating. Second,
as the cranial base articulates with almost all the structures changing during growth and evolution, its summary in a few discrete sta-

tistics is problematic. It is not just the kyphosis which matters, but also how the other structures adjust their own positions to it in the light of their own functional needs. Moore and Lavelle (1974), Gould (1977: ch. 10), and Scott (1958) all attempt a synthesis of the empirical literature, but it seems impossible. The available measurements do not support a clear exegesis. This is, I think, due to a general morphometric disability (see chapter iii) to which the cranial base contributes no special complications.

There are two basic difficulties in such expositions: the lack of an integrated statistical method corresponding to the wholeness of the form, in principle the sole valid object of study; and the inability to take ontogeny into account other than by controlling for developmental stage. Never have both of these problems been ameliorated simultaneously. In one approach, a set of forms is stereotyped in a conventional diagram of landmarks and segments. The plain visual contrasts among these graphs give rise to angles and ratios which are "interpreted" or summarized in a mass of tables and graphs. Here the diagrams provide rather more memorable summaries than the statistics which purport to dissect them. Thus Biegert (1957), considering adult and juvenile forms over a wide range of genera, sketches in pen the relative positions of several cranial structures, then measures them arbitrarily but redundantly enough to typify all sorts of trends which were already jointly apparent. Verheyen (1962), Vogel (1968), and Dmoch (1975-6) reduce their data to sequences of variously complicated polygonal outlines which are both displayed and atomised into lists of means over populations. The statistics of such essays are less definitive than the diagrammatic data deserve, for graphical nuances of correlated changes are very difficult to express in relations of indices separated on the skull. Works which are not fundamentally graphic but attempt nevertheless to compare ontogenies are even more frustrating: cf. Moore and Lavelle (1974), in which one can simply not see what is going on.

In researching this subject, one sporadically encounters the holistic description of change even in the absence of suitable quantification. Aside from the elegant superpositions of Delattre and Fenart (1960), all these studies use the familiar device of the Cartesian grid. I have located depictions of ontogeny (Starck and Kummer, 1962; Kummer, 1953), adaptive radiation (Kingdon, 1971:1:143, 147), and phylogeny. This last category Thompson himself initiated (cf. Fig. V-3), and many later workers have attempted to improve his images or apply them to fossil data. Figure VII-2 is a sample from

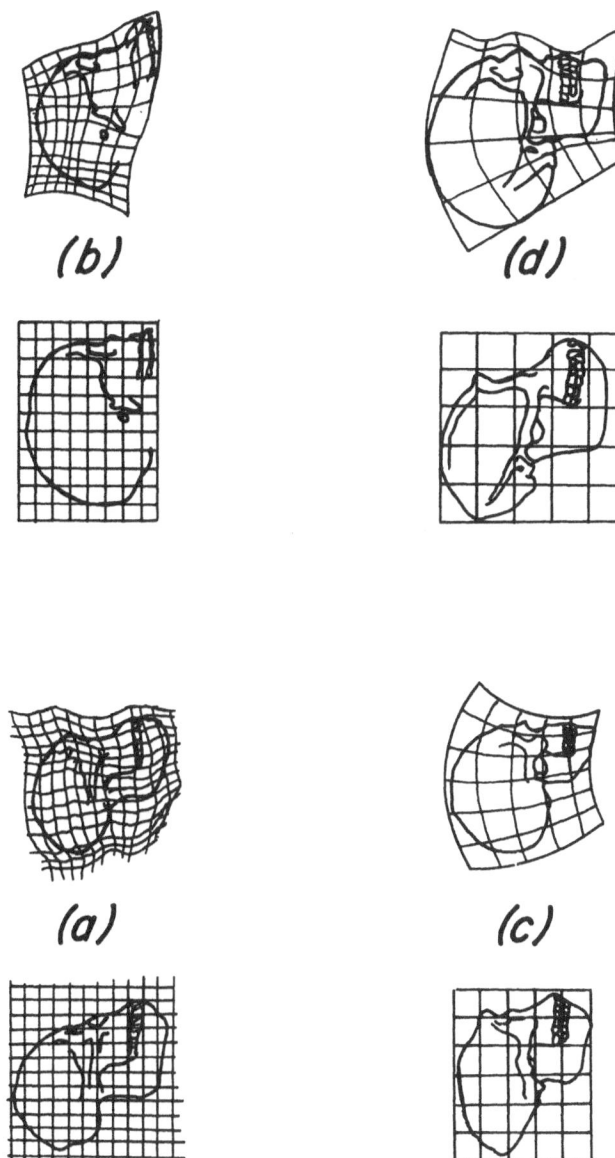

Fig. VII-2.--Various grid drawings of hominization, all in
Thompson's style. (a) From Proconsul to Australopithecus, after Appleby and Jones (1976). (b) From Homo sapiens to Australopithecus,
after Sneath (1967). (c) From Pithecanthropus robustus to H.
sapiens, after Hutchinson (1948). (d) From Proconsul to H. sapiens,
after Kummer (1952).

this congeries. It is not clear how to compare these among them-
selves--it is not even clear whether they grossly agree.

Without a technique for extracting quantitative features, the
several diagrams of earlier publications are but muffled (although
pretty) expressions of complex trends which must be reported by other,
non-pictorial means. One must properly separate the contributions
of ontogeny and phylogeny to human cranial morphology--examine at
once the pattern of change over the life-cycle and over the eons in
comparative quantification. The biorthogonal method is ideal for
this task. It is fundamentally graphic, and can summarize form
changes without their prior reduction to indices; but then the transforms themselves may be contrasted among sets of three or more forms
by the machinery of the grid. We thus directly compare the effects
of ontogeny and phylogeny upon the same form, disentangling the con-
foundment inherent in the conventional metric literature by independently extracting features of the shape changes separately.

2. Data for this exercise

The geometric object of study is the form of the cranium (the
skull, excluding the mandible). We study it in sections of the head
through its single plane of symmetry: <u>midsagittal</u> sections.

Photographs of the midsagittal plane itself in hemisected animals, with bony textures and space-filling organs in great variety,
can be found in Hofer (1965 and references therein). For the biorthogonal analysis this complexity must be abstracted into outlines,
so that homology can be seen as an idealized smooth correspondence
of loci inside idealized homologous boundaries. According to the
principles of Part One, we replace the raw data with sets of landmarks connected by smooth curves. In the series of diagrams which
follows, landmarks are indicated by an X on the outline. Clockwise
from lower left, they are: prosthion, the frontmost point of the
alveolar margin of the maxilla; rhinion, the bottom of the internasal suture; nasion, the top of that suture; inion, center of the
bump where the nuchal torus crosses the midplane; basion, frontmost
point of the foramen magnum; and an unnamed point slightly toward
the chin from staphylion (back of the bony palate) where the last
molar erupts from the alveolar bone. In addition, there is one interior landmark, sella, drawn as a square dot near the lower central
margin of all forms. These landmarks are strictly upon the midline
except the penultimate, unnamed one on the molar. I prefer it to

staphylion, which is on the midplane, because the molar locus represents better the dentition function within the oral cavity. Several
other interesting landmarks, associated with the ears, the orbits,
and the zygomatic arches, cannot be used here, for reason of their
great distance from the midplane.

The data represent the cranial base not by an extended structure but by its "center" (sella) and approximate endpoints (basion
and nasion) unconnected. The use of nasion in this regard is not
without misgivings. Several authors, including Scott (1958) and Moss
(in Salzmann, 1961:47) point out that the coupling between nasion
and cranial base is uncomfortably loose: the former continues to move
in ontogeny long after the latter has ceased growth. I use nasion
in this study nevertheless, as all alternatives lie well inside the
outline.

The arcs of the outlines are as follows. From prosthion to
rhinion a straight segment bounds the "empty space" (in this section)
of the nasal aperture. The internasal suture above it is modelled
simply as the arc of a circle. From nasion to inion the outline is
a single conic arc (usually an ellipse, but not always) fitted through
bregma and two additional points on the outline. This arc, passing
inside glabella and external inion, systematically fails to represent
any brow ridge and nuchal torus which might be present, for these
are only local structures. Bregma does not have landmark status
in this study. It appears to move about the vault independently
and capriciously over phylogeny, an accidental consequence of a badly-
regulated sequence of fusion of ossification centers, and further
is difficult to spot in older specimens whose coronal sutures have
closed. From inion over to basion the outline is a circular arc
through opisthion, so that the foramen magnum is in fact a chord of
this arc, of one endpoint indeterminate. The line segment from basion to alveolar ridge cuts across empty space outside the cranium,
underneath the pterygoid plate; the true form here is too complex
for any sort of smooth rendition. Finally, along the alveolar ridge
the outline follows a circular arc through an intermediate point
at the gum line. The sella is inside this form, more or less deeply
as the cranial base is more or less bent and the occlusal plane more
or less lowered below it. Horizontal is set parallel to a segment
from basion to nasion, after the fashion of Schultz (1955). As the
biorthogonal method is quite independent of orientation, this stan-
dardization is purely for the convenience of the reader.

The outlines were drawn using digitizations of the forms in

Biegert (1957) and Abbie (1963). These are drawings of actual mid-
sagittally sectioned surfaces, so that none of the adults are the same
individuals as those figured in infancy. Then all computed ontogenies are approximate. More suitable raw data--accurate tracings of
midsagittal sections in longitudinal series--are unexpectedly scarce,
perhaps because their analysis by existing methods is so equivocal.
Apart from Biegert's portfolio, the only reliable sources are the
articles of Hofer cited in Hofer (1965). Hershkovitz (1977) presents
a good assortment of adult forms, but no juveniles.

The style of these outlines improves that of my previously published data analysis, Bookstein (1977), in several respects. Here
the external auditory meatus is replaced with basion, truly on the
midplane; here sella serves as milestone on the path of the cranial
base. The new analysis uses empirically curved arcs along the vault
instead of chords cutting across large bulges, and it abandons the
bregma as a landmark, presuming instead that homologies of ontogeny
or phylogeny are linear over the whole vault from nasion to inion.
The study of ontogeny, of course, does not extend to human paleontology, so I have not digitized any hominid fossil forms here.

All transformation grids in the following figures are computed
according to the algebra of Technical Note 2. Boundaries are mapped
onto boundaries by homologies linear in arc-length between landmarks,
with an appropriate correction so that sella maps onto sella.

3. Two types of transformations

By way of introducing the analysis of these grids, I shall display two extreme cranial forms for which ontogeny closely parallels
phylogeny.

In the howler monkey, Alouatta, the mandible is immensely enlarged to make room for specialized vocal apparatus. This distension
alters the form of the basal cebid stock (cf. Biegert, 1957): it
tilts the maxilla up and forward out of the way, moves the foramen
magnum backward, and affects proportions everywhere in the skull.
Figure VII-3(a) displays this alteration using schematics of cebus
and howler outlines as previously described. The primary pattern of
biorthogonal curves is quite striking: an expansion along radii about
a point somewhere above the clivus. Linear extension away from this
point increases in rate along rays as they approach the lower border
of the form, the whole of which is being extended so; but the flattening of the vault and face and the rotation backward of the foramen
magnum are all geometrically coordinated with this extension accord-

ing to the strains as drawn. For instance, the opening of the fora-
minal angle is here expressed as the greater extension backward of
its lower border than its upper, a local inequality which persists
throughout the length of the head.

Exactly this same system of axes and dilatation gradients can
be found in the ontogeny of the howler itself, Fig. VII-3(b), for,

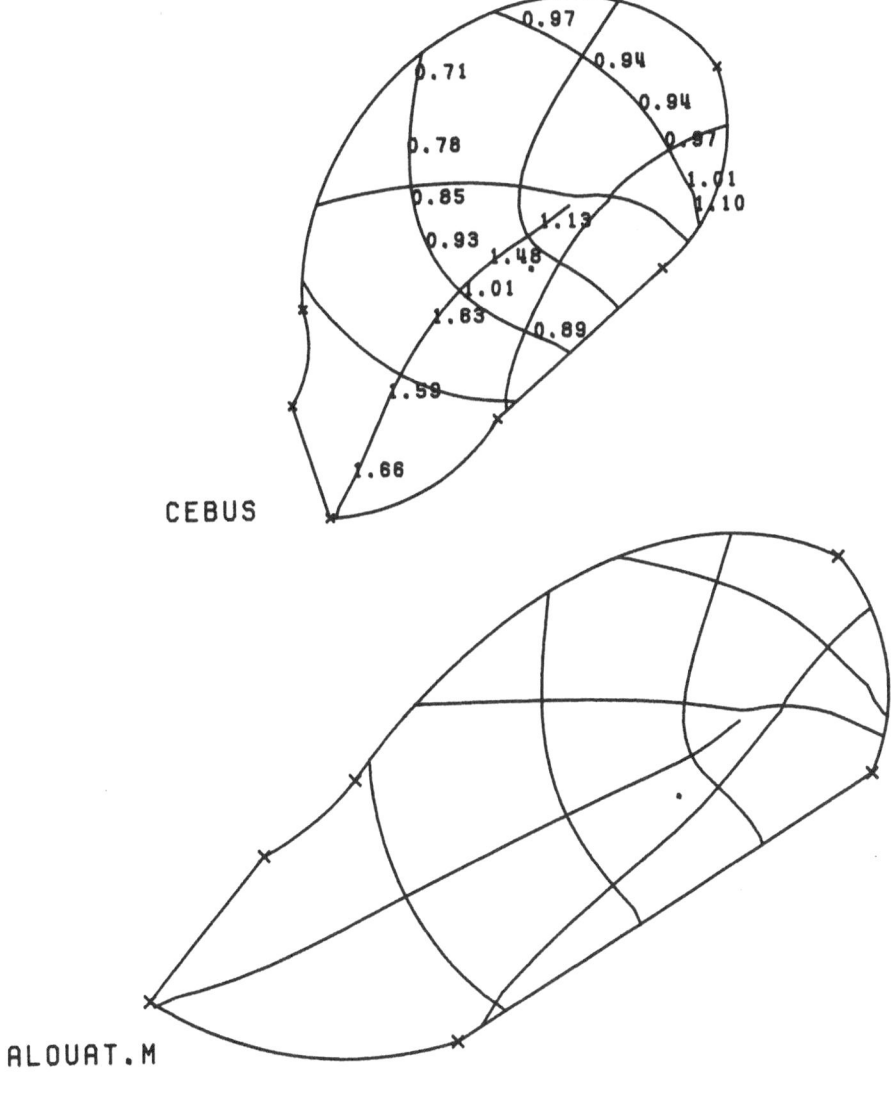

Fig. VII-3(a).--Phylogeny of the howler monkey.

as it happens, the infant howler looks identical with the cebid adult
in this diagramming technique. The main feature of the ontogeny is
a general stretch in all forward directions away from a point mid-
foramen. Along these directions, growth is distension by a factor
greater than two at and below sella. Perpendicular to these curves,
weak gradients fail to preserve the infantile rounding of the calva,

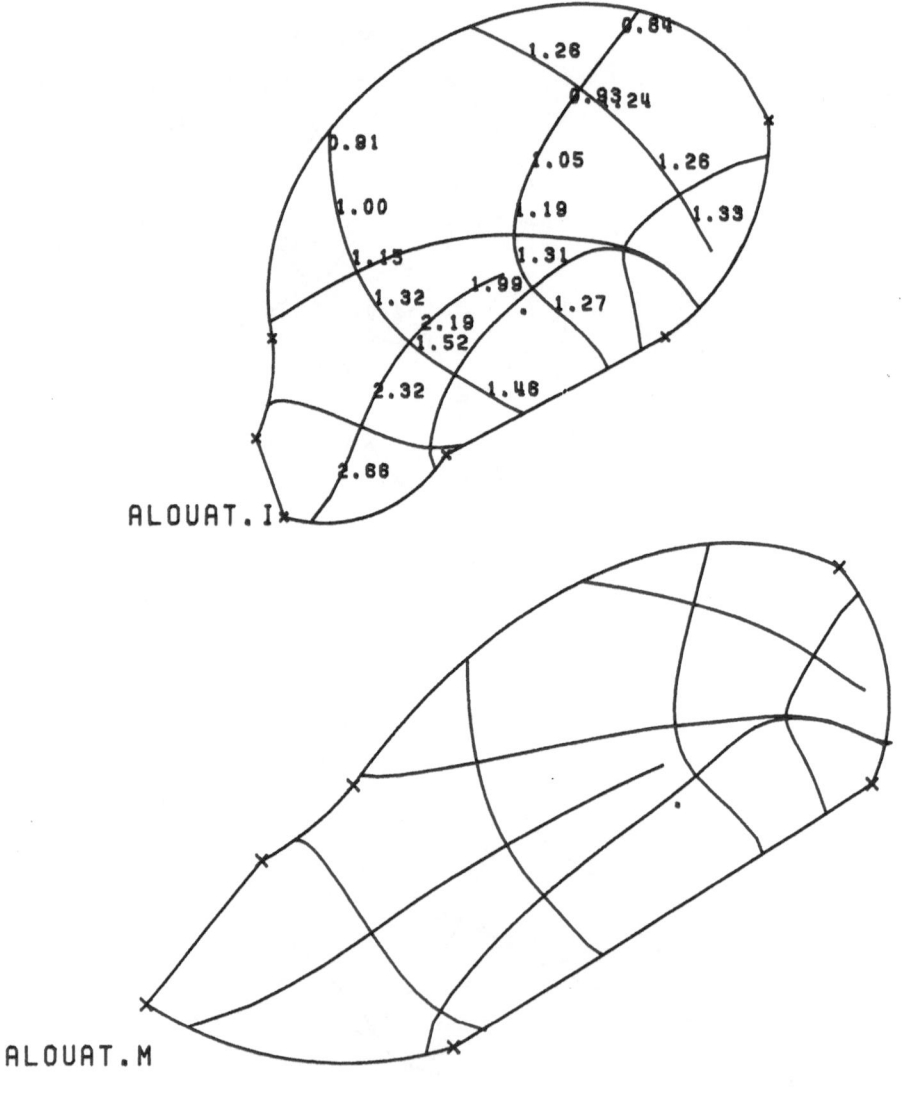

Fig. VII-3(b).--Ontogeny of the howler monkey.

which is expanding near the cranial base faster than near the crown
in both horizontal and vertical directions.   These axes then represent a sort of anti-cephalization by growth of nearly the whole articulation with the mandible, not of the jaws or the cranial base
specifically.   Closely homologous axes and gradients can be noted also
in the ontogeny of the female howler and, to a good approximation, in

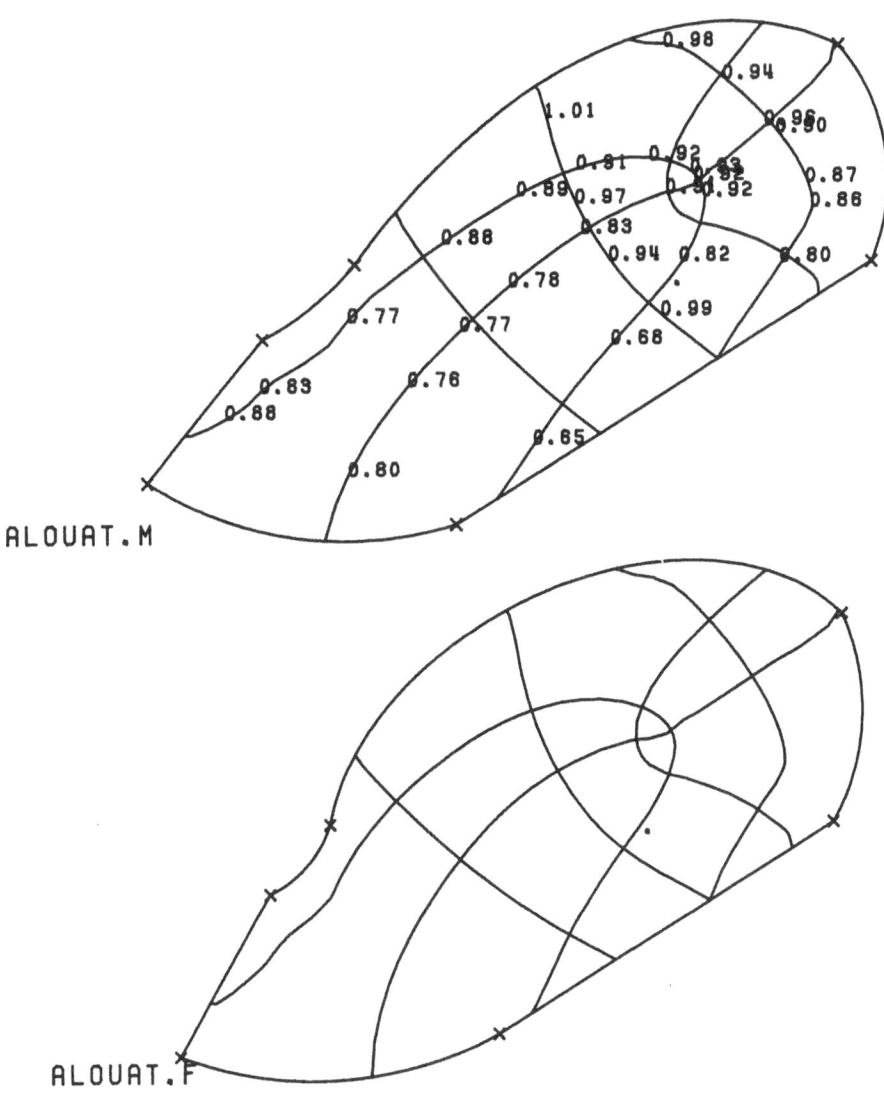

Fig. VII-4.--Comparison of the sexes of the howler monkey.

the comparison between the sexes, Fig. VII-4. One may infer a characterization of the sequence infant-female-male as a single structure of deformation, followed by its intensification, exactly as in
Thompson's discussion of human, chimpanzee, and baboon, page 71.
Here, then, ontogeny "recapitulates" phylogeny quite clearly. A
single forcing function, vocalization, arises rather late in ontogeny and manifests itself in systematic and consistent distortion

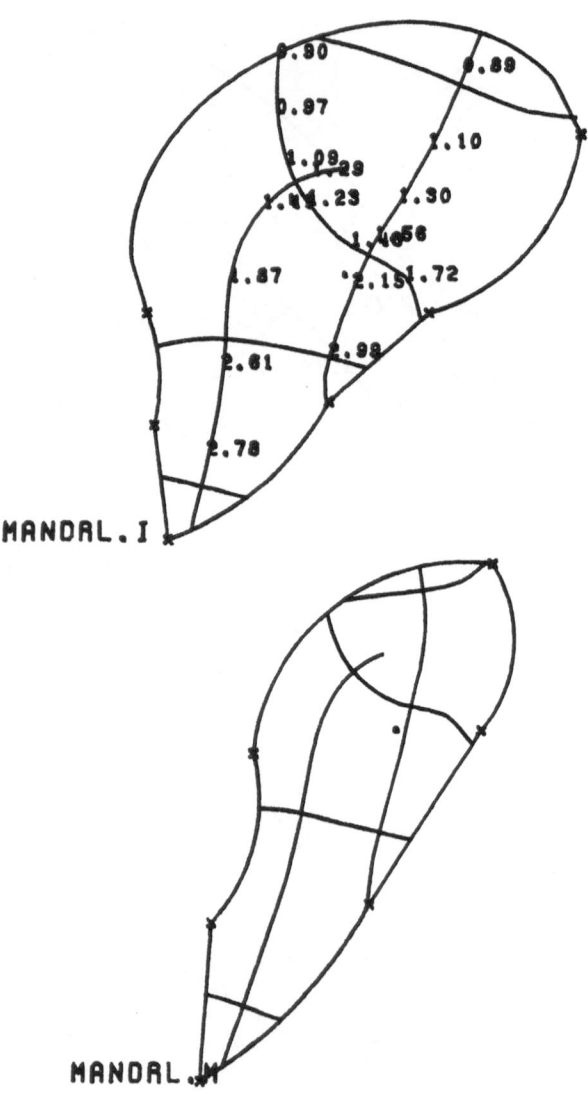

Fig. VII-5.--Ontogeny of the mandrill.

varying only in magnitude between the sexes.

Another instance of the same consistency appears in the well-
known allometric size sequence of the cynopithecines, from macaque
through gelada and baboon to the mandrill. In the ontogeny of the
mandrill, Fig. VII-5, there is a stretch of the jaws aligned with
the face (not along the base of the form as for <u>Alouatta</u>). The

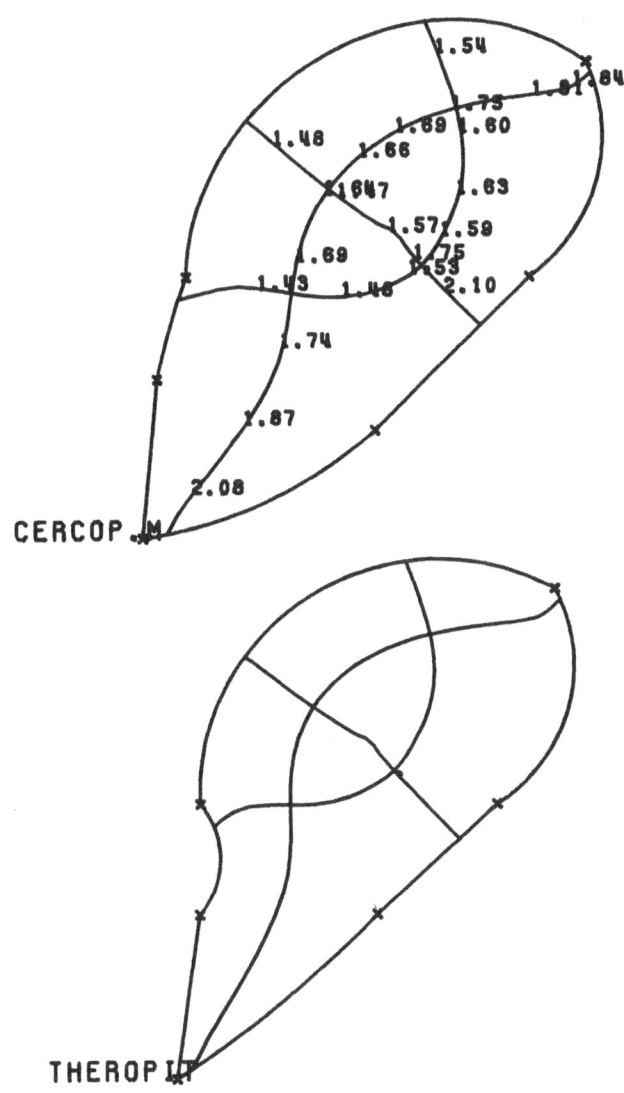

Fig. VII-6(a).--Comparison of guenon with gelada.

whole muzzle has grown disproportionately, in gradients increasing toward the dentition; as the calva fails to increase much in height, there appears a considerable flattening; the foramen magnum rotates, but to a lesser extent than for howlers. Now the mandrill is largest of the cynopithecines, and its ontogeny is quite consistent with a hypothesis of phylogenetic shape intensification as adjustment from

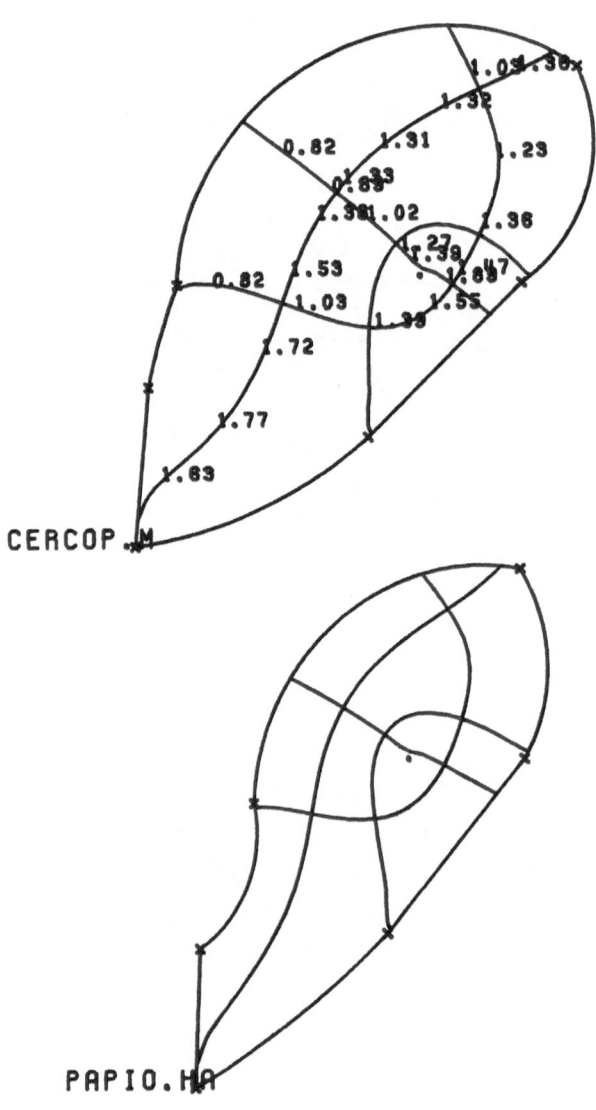

Fig. VII-6(b).--Comparison of guenon with hamadryas baboon.

the basal cercopithecoid stock. Figures VII-6 (a,b,c) show the biorthogonal grids for comparison of a male guenon with gelada, hamadryas baboon, and mandrill, respectively. In each there seem to be
two main systems of axes, one running the length of the head, the
other its height from sella to the vault. In the upper engridments
the stability of the sequence of transformations can be seen quite

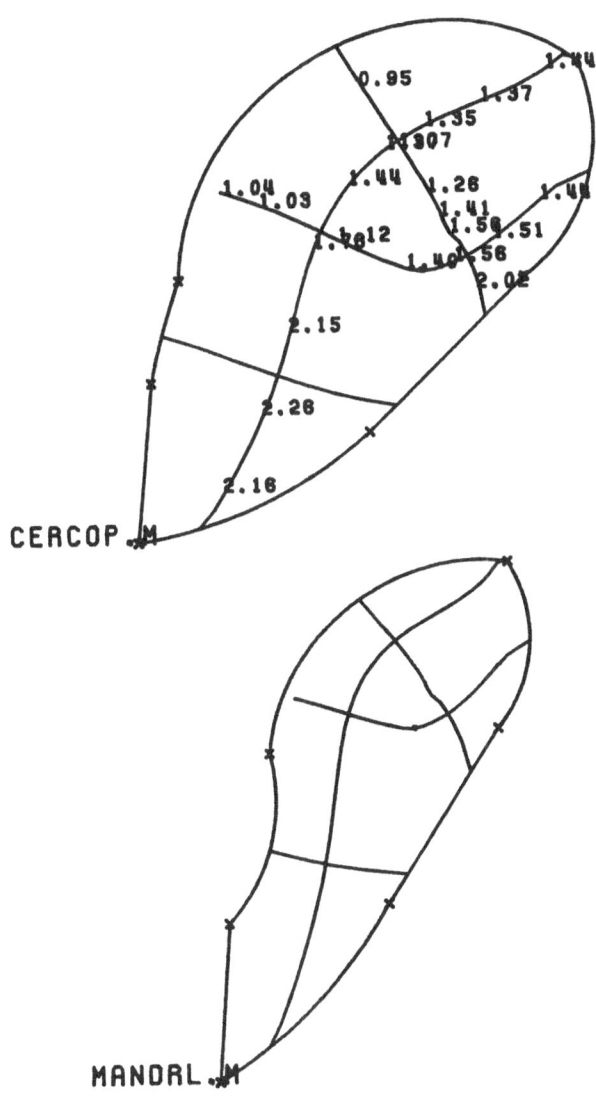

Fig. VII-6(c).--Comparison of guenon with mandrill.

clearly, and also the intensification of gradients along the two axes. The mandrill ontogeny strongly corroborates this. By its quotation of the gradients of the cynopithecine sequence it appears to be the top end of an adaptive radiation by intensification of a single system. The modification of ontogeny which the size-sequence entails is not here postponed as for the howler: there is a great divergence between the adult guenon, Fig. VII-6, and the infant mandrill, Fig. VII-5.

Consider the stylized hieroglyphic eye in these last figures, delimited by coordinate curves which bend around a center to intersect each other twice. Inside is a reliably occurring singularity of the biorthogonal coordinates. The main length-gradient of the head--growth at the jaw, more nearly stasis near the inion--is itself graded, appearing somewhat lesser near the frontal, somewhat greater near the cranial base. Perpendicular to this gradient, as seen in the figures, is a gradient also increasing from calva to sella. These together engender a singularity somewhere at which the gradients are exactly equal in all directions, so that the axes technically cannot be computed. Around this point the biorthogonal curves are diverted as shown--these singularities all appear to be of the one-root form described in the previous chapter's Technical Note 1. The presence of the singularity is a function of the relative magnitude and location of those gradients along length and height. Whether they cross or not, they embody the opening of the cranial base angle, that hard-to-measure kyphosis, in the excess of horizontal dilatation over vertical inside the kyphosis, along the front underside of the schematic form. Where the biorthogonal axes are nearly at 45° to the aperture of this angle along the path of the cranial base, the kyphosis has changed hardly at all; where they are aligned with the aperture of this angle, the kyphosis has changed a great deal, having been "forced" open or "pulled" shut (in the metaphors of chapter vi).

4. Comparative ontogeny of the apes and man

With these two somewhat extreme examples in mind, representing two distinct varieties of "anticephalization," we may now consider the relatively most cephalized primates, the apes and man. Figure VII-7 presents the biorthogonal grids for the infancies at opposite ends of the main size sequence, of gibbon versus gorilla. As we saw in the cynopithecine radiation, there are two major gradients, one the length of the head, one up from the cranial base, which cross somewhere near sella. Comparison of infant gibbon with infant gorilla

indicates a lengthening of the latter with respect to the former
which is faster than increase in height at the vault, but slower than
increase in height near the sphenoid. The lower half of the infant
gorilla form, in short, is deepened and lengthened in comparison
with the top, in anticipation of the major expansion of the jaw and
its articulations which is yet to come. The same can be seen in the

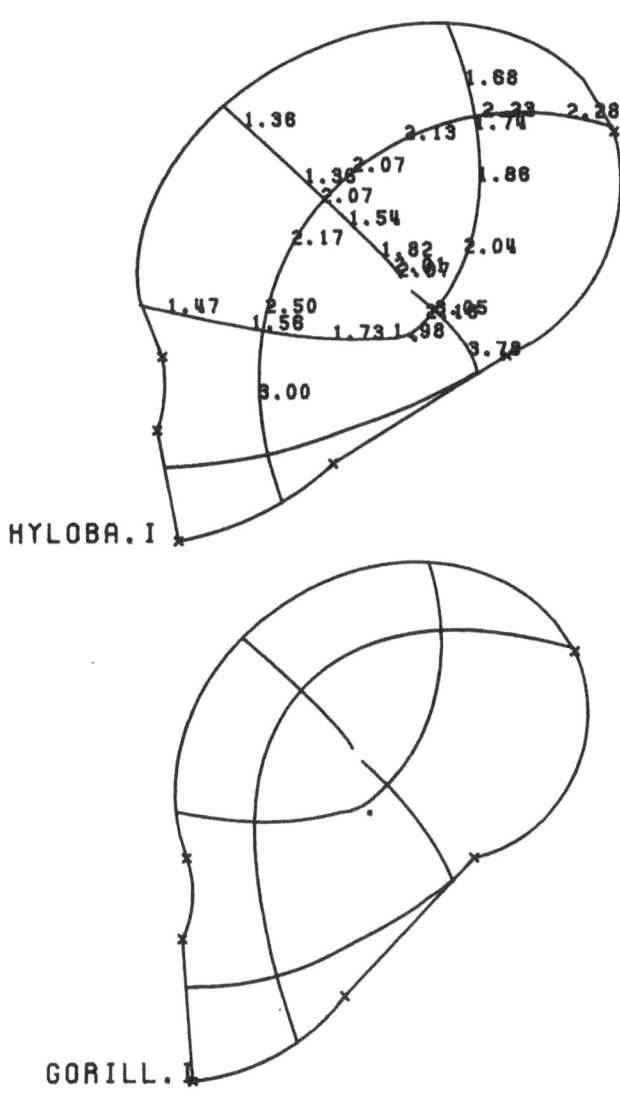

Fig. VII-7.--Comparison of infant forms of gibbon and gorilla.

comparison of the infant chimp with the gorilla, the infant gibbon
with the orang, etc.

The general trend of ontogeny among the apes is not aligned with
this general pattern, but rather manifests a system of gradients without singularity. Figure VII-8(a) is the Cartesian grid for the
growth of the male gorilla. (The sagittal crest lies well outside

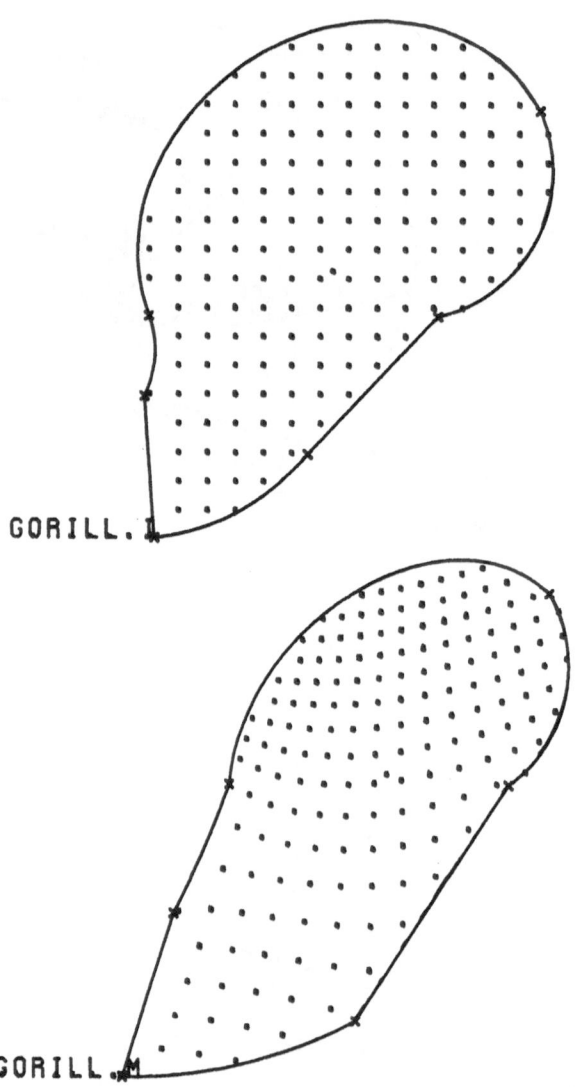

Fig. VII-8(a).--Cartesian depiction of male gorilla ontogeny.

the form as drawn: the arc along the vault of the skull proceeds
parallel to the inner table at fixed distance outside it.) This
Thompson-style grid is almost readable by itself, as, by accident of
orientation and deformation, the vertical straight lines of the engridment upon the infant form are mapped very nearly into straight

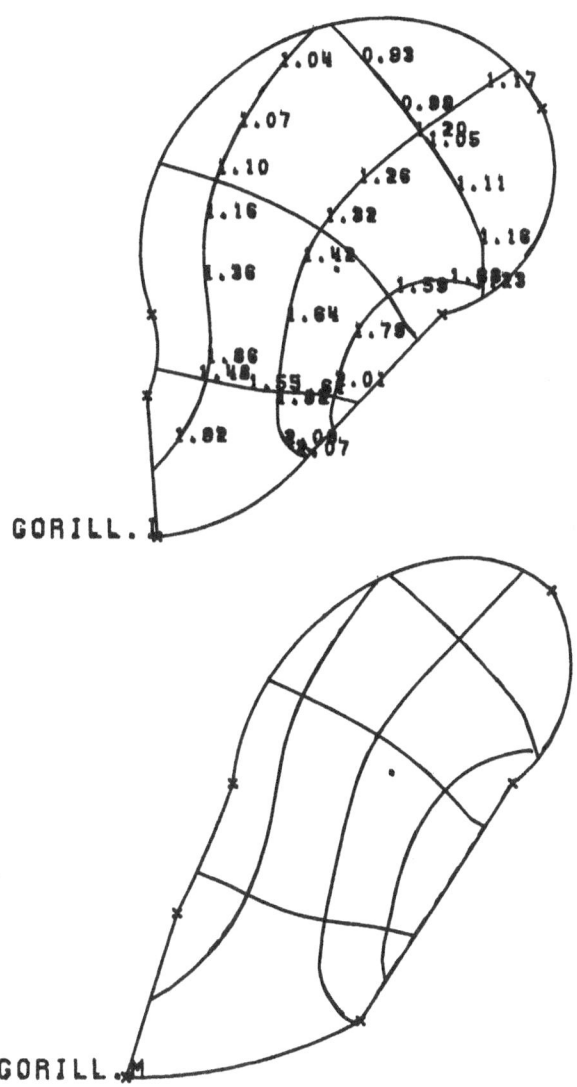

Fig. VII-8(b).--Biorthogonal depiction of male gorilla ontogeny.

lines upon the adult form. They suggest a radial stretch and an
opening-out of the angle about acrion. Figure VII-8(b) makes this
impression precise. It shows a pair of major gradient systems of
which the horizontal is always more pregnant than the vertical. The
opening of the sellar kyphosis is plain here, as the facial skeleton
is enlarged "forward" much faster than it is moved down. The biorthogonal grid also shows clearly the forward repositioning of the
maxilla in the course of its massive expansion. The gradients of
dilatation are consistent with those for the other ontogenies we
have seen, and represent, again, an anticephalization: length
stretch increases downward from the inion, while height stretch increases downward toward the cranial base. Comparison of the sexes
shows, as for _Alouatta_, that the female ontogeny embodies the same
gradients as the male's, but slightly weakened in all respects. For
the orang, the other great ape, both the growth grids and the contrast between the adult sexes are quite similar.

   Examine, now, the relation between infant chimpanzee and infant
human, Fig. VII-9(a). This grid is not of the same family as the
main-line infant sequence, Fig. VII-7, but rather looks like the pattern we saw before in _Alouatta_, only in reverse. The larger form,
Homo, is here shown to be _unintensified_. The cranial base and the
tooth row have enlarged more slowly than the vectors perpendicular
to them, and nasion moves away from basion faster than from prosthion, indicating a system of growth-gradients radiating outward
from somewhere near the inion; the kyphosis angle has somewhat closed.
The comparison of adult forms of the chimp and man, Fig. VII-9(b),
looks remarkably like the comparison of their infancies, and shows
a further retardation of the human form, a further decline of the
maxillary apparatus and sphenoid relative to the vault in both height
and length.

   The _ontogenies_ of chimp and man are likewise similar (Figs.
VII-10(a,b)), not to the ontogenies of the other great apes but to
the relations among their infant forms. Growth in length dominates
growth in height at the top of the head but is dominated by it at
the base: this is the infant series Fig. VII-7, not the consistent
dominance of length over height in the ape ontogeny of Fig. VII-8.

   This little investigation into the form-changes of the anthropoid cranium in mid sagittal section speaks directly to the old
controversy about the "fetalization," the neoteny of man. Judging
by these diagrams, which express the coordination and geometric relation of growth gradients and size contrasts all over the skull, it

is erroneous to speak of the neoteny of man in isolation. Rather, the chimp and man manifest similar gradients in ontogeny, so that the relation of their infant forms is concordant with the relation of their adult forms, indicating an intensification of the retardation of the chimp with respect to the main anthropoid line. Both their ontogenies are unlike the ontogenies of the other apes, in fact resembling the comparisons among the infant stages of the latter. In each form, the lower margin of the cranium--the curve from basion to prosthion--is disproportionately small for the beast's overall size, in an exact reversal of the Alouatta idiosyncrasy. Indeed Homo manifests this effect more intensively than Pan, and this is apparent from the earliest stages in this data set. The discrete shift, the neotenization, appears in common between Homo and Pan, and is merely intensified in the former ontogeny in comparison with the latter.

As I indicated at the outset, to verify this argument properly requires, besides the computer program I have demonstrated, two addenda. The data must be collected with this hypothesis in mind: true ontogenetic series from longitudinal x-ray studies. The forms involved must not be ideal types, but instead representative of population variations in both ontogeny and final form. Second, the presentation here is in essence a factor analysis of the gradients I have been showing. I have tried to sort the comparisons into two distinctive geometries: the infant series, with a singularity, as also seen in the cynopithecines; and the hominoid series, without, showing the Alouatta-like frontal stretch. Factoring tensor fields is beyond my algorithms at the moment, so I could only hint that the deviation of the human adult form is a mixed consequence of its intensification of the retarded chimp infancy and of the joint deviation of their ontogeny from the norm. In exchange, we have at least avoided the bootless argument over measurement of cranial base flexure, prognathism, and all the other separate features of the skull in section. The biorthogonal method generates the proper features automatically.

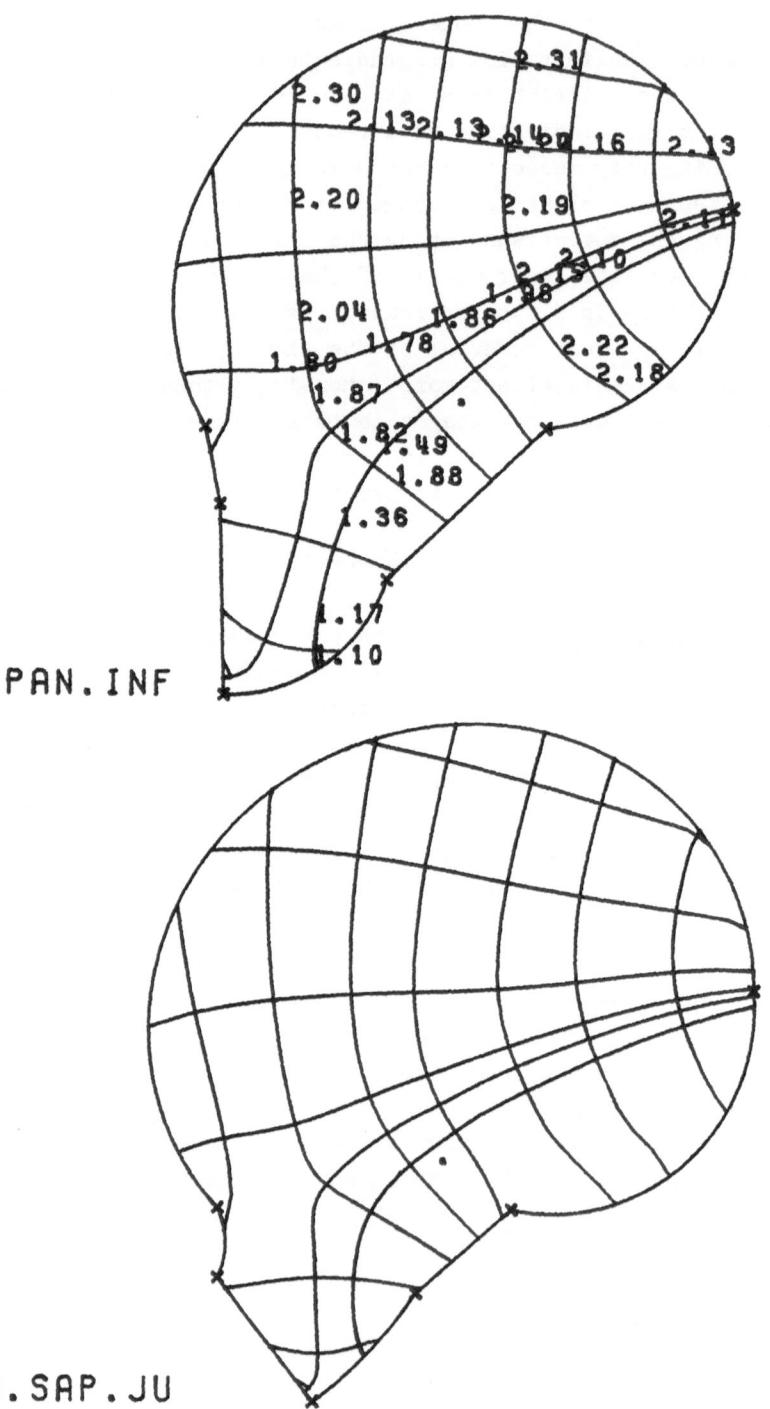

PAN.INF

H.SAP.JU

Fig. VII-9(a).--Comparison of infant forms, chimpanzee and man.

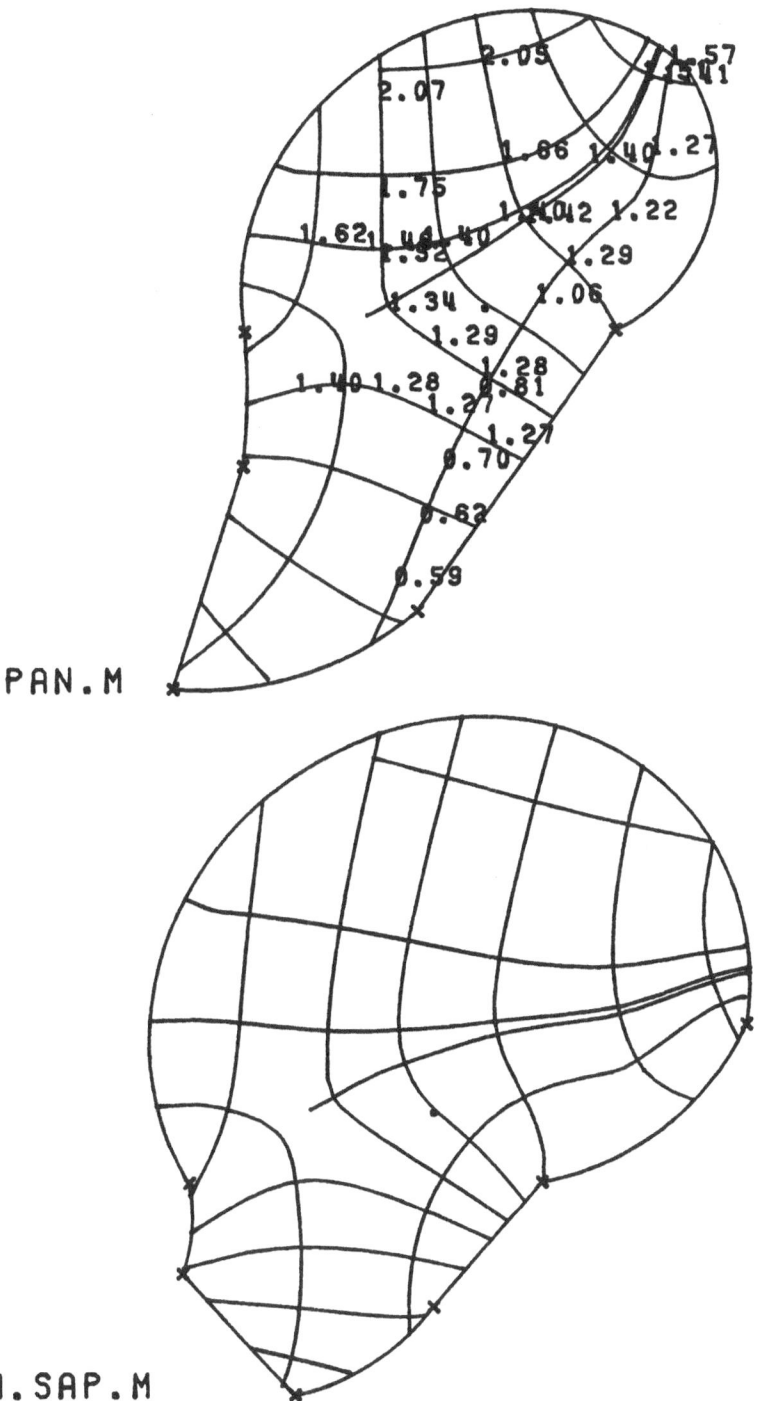

PAN.M

H.SAP.M

Fig. VII-9(b).--Comparison of adult male forms, chimpanzee and man.

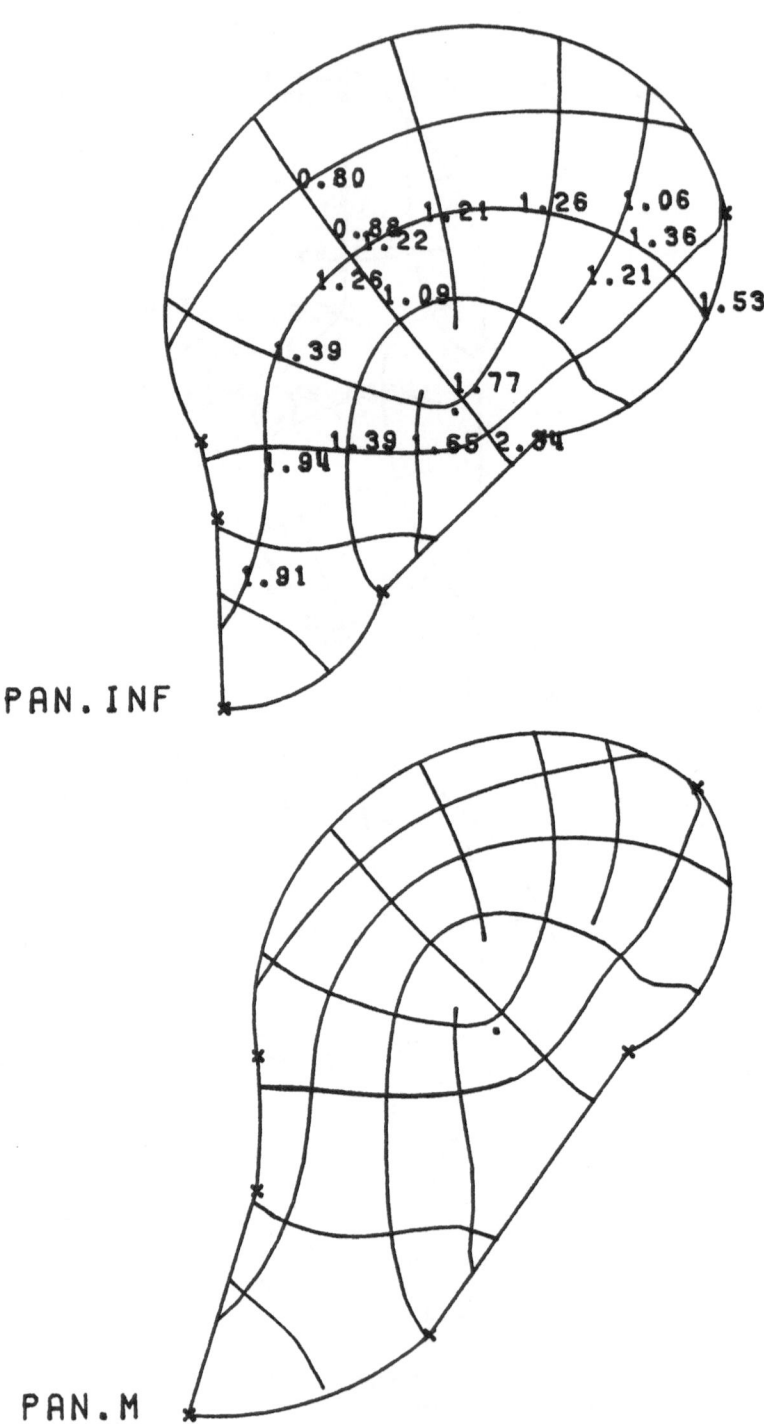

PAN.INF

PAN.M

Fig. VII-10(a).--Ontogeny of the male chimpanzee.

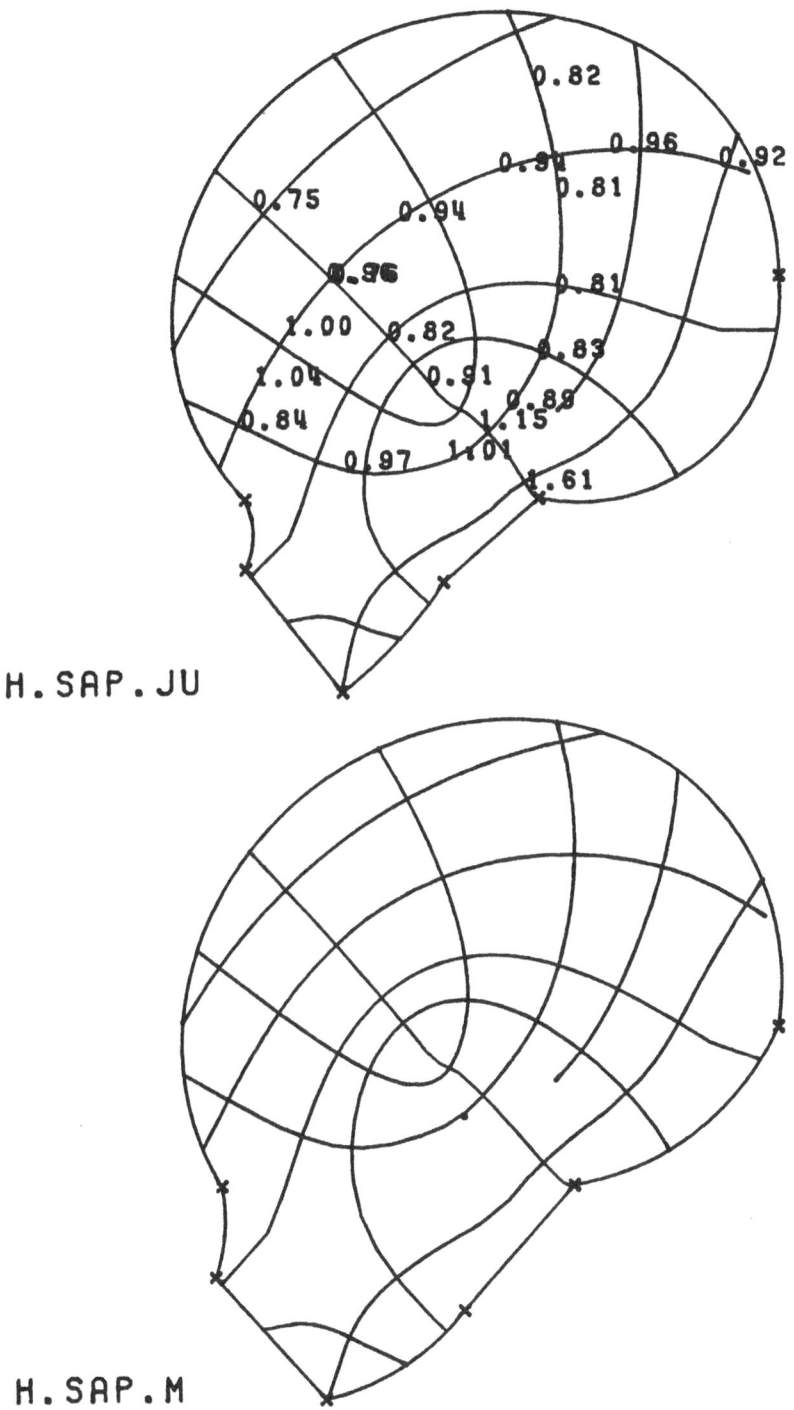

H.SAP.JU

H.SAP.M

Fig. VII-10(b).--Ontogeny of the male human.

I believe the method of biorthogonal grids to be not a compromise
with quantification, like those of chapter v, but a basic step toward
the direct and automatic quantification of transformations. The only
test is by empirical exercise, in the analysis of sequences for seriation or prediction and in the statistical and substantive interpretation of the dilatation patterns extracted. Of course, a great many
methodological questions remain to be explored. Three classes are
particularly crucial: statistics, computations, and interfacing with
various applied problems.

A. Statistical Methods

The symmetric tensor field. The representation of a Thompson
shape transform is in effect by its matrix derivative expressing the
infinitesimal affine transformation at every point of one of the
shapes, as described in Technical Note 1. This matrix anywhere can
be written in the canonical form $O_2(x,y)D(x,y)O_1(x,y)$, where (x,y)
is the variable point at which we have taken the derivative, O_1, O_2
are orthogonal matrices (rotations), and D is a diagonal matrix bearing the dilatations. This decomposition of the Jacobian is known
variously as the polar form (Jacobson, 1953: ch. 6, theorems 4, 12)
and the singular-value decomposition (Golub and Reinsch, 1970).

The pieces of this decomposition correspond to successive steps
in the following model of the transformation in a little neighborhood of (x,y):

$$O_1(x,y) \qquad D(x,y) \qquad O_2(x,y)$$

Now O_2, the reorientation of the axes in the image shape, is but a
consequence of pushes and pulls elsewhere in the organism. Given
only the fields O_1 and D, the transformation is known in detail up
to an inessential rotation of the whole second image. The irrele-
vance of O_2 encapsulates the fact, alluded to earlier, that local
neighborhoods do not know anything of rotation, only of differential
dilatation. With a single $O_2(x,y)$ fixed for any arbitrary point
(x,y), O_2 for all the others follows from the implications of the
differential growth upon integration.

Then we are free to register the shape change in any fashion
we wish, and it is convenient to let the registration vary with the
point (x,y) in such a way that the canonical axes are parallel in the
two images. Such a constraint requires $O_2 = O_1^{-1}$, the opposite rotation. The picture becomes:

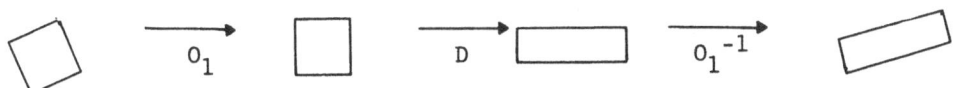

Modelling the derivative by a matrix of the form $O^{-1}DO$ everywhere
frees us from the accidental consequences of growth elsewhere and
reduces the system to three local parameters. The matrix O tells
us of the orientation of the axes and the elements of D are the dilatations. This gives us all the information we need to reconstruct
the change, once one O_2 anywhere is arbitrarily set, for all others
follow from the implications of the differential growth upon integration.

The matrix $O^{-1}DO$ is symmetric. It may be thought of as the
positive-definite square root of AA^t, where t is the transpose operator and A is the original derivative O_2DO_1, or as the Gauss-Riemann
metric tensor giving distance in the second figure in terms of coordinate differentials in the first figure: $ds_2 = \sqrt{(dx_1, dy_1)O^{-1}DO}$
$(dx_1, dy_1)^t$. In either interpretation, shape change is embodied in
a symmetric second-order tensor field.

The operators representing shape change are thus reducible to
fields of symmetric matrices on the points of the "ground" figure.
We can multiply or divide any two of them, pointwise, and get another
such; for at every point the product or quotient of two positive-
definite symmetric matrices is likewise positive-definite symmetric.
It would be possible to analyze a complete experimental design of
shape transforms by breaking them down into products of components
just as anova breaks down means into effects; we might emerge thereby
with models for the joint action of genes or morphogenetic gradients
or simultaneous distributed causes upon biological shape. More directly to Thompson's purposes, we can raise a shape change to a power,
a fractional iterate. The eth power of a change whose tensor field
is $O(x,y)D(x,y)O^{-1}(x,y)$ is the field $O(x,y)D^e(x,y)O^{-1}(x,y)$, where if
$D = diag(d_1, d_2)$ then $D^e = diag(d_1^e, d_2^e)$. This corresponds point for
point to the method I discussed previously for iteration of affine
transformations.

Concordance. Three shapes A, B, C in perfect series will have
growth fields which are perfect powers of one another whenever we
represent them upon the same ground image. This implies concordant
rotations O everywhere and the existence of scalars s, p (scale and
power, respectively) such that $D_{A-C}(x,y) = sD_{A-B}^P(x,y)$. Note the
resemblance of this form to the allometric equation $y = ax^b$. If
this is the case for the ground figure A it is the case for any of
the others.

To compute s, p for concordant affine transformations we write
their derivatives, here presumed independent of position, in their
polar forms $O^{-1}D_1O$, $O^{-1}D_2O$ where $D_i = \text{diag}(a_i,b_i)$, i=1, 2. Then
$p = \log(a_2/b_2)/\log(a_1/b_1)$ and $s = a_2/a_1^P = b_2/b_1^P$. To compute s,
p for two general transforms, not necessarily affine, one might estimate their values globally by some sort of least-squares method. I
suggest instead the following strategy. The quantities s and p can
be computed as scalar fields $s(x,y)$, $p(x,y)$ wherever the local canonical axes are not too badly misaligned, by projecting the observed
dilatations to new axes midway between the two slightly discordant
original sets. If the angle between the axes (i.e., the net rotation
$O_2O_1^{-1}$ at (x,y) between the polar forms of two transforms $A_i = O_i^{-1}$
$\text{diag}(a_i,b_i)O_i$, i=1, 2) is 2θ, replace a_1 by $a_1' = \sqrt{(a_1^2\cos^2\theta + b_1^2}$
$\sin^2\theta)$ and b_1 by $b_1' = \sqrt{(a_1^2\sin^2\theta + b_1^2\cos^2\theta)}$, and similarly a_2, b_2.
These are point parameters; we can compute them throughout suitable
concordant regions all over the ground image. They have no clear
numerical equivalent in the allometric tradition, where one always
requires two anatomical points for a coefficient and three points
for a growth-gradient; in contrast, here we require two images for
a coefficient (the dilatation matrix) and three images for a gradient
(the s, p under discussion), but only in a single infinitesimal neighborhood of one point. The $s(x,y)$, $p(x,y)$ extracted are ordinary
scalar fields which may be inspected for regional variations, fitted
to trends, smoothed, and the like. Values of s and p for different
regions may be compared; when one p is double another, directional
growth is proceeding twice as fast. We may set models of constant p
(or s), and compute the global p (or s) of best fit according to
different models of regional variation in s (or p); there will then
appear "residual dilatations" between predicted and observed growth,
pointwise.

At the same time, a third scalar field can represent the mis-

alignment itself. The angular disagreement of two affine derivatives
is most easily expressed by a simple trigonometric function of the
angle "between" their symmetrized forms $A_i(x,y) = O_i^{-1}(x,y)D_i(x,y)$
$O_i(x,y)$, i=1, 2. A suitable expression is $|\sin 2\beta(x,y)|$ where $\beta(x,y)$
is the angle between any axis of one canonical set and any axis of
the other. The pairs are minimally discordant when they line up
perfectly, maximally when each one bisects the other's right angles
$(\beta = \pi/4)$.

This simple measure is qualitatively satisfactory, but it does
not quite capture the exact information we need about orientation.
Where the two derivatives A_1, A_2 are highly anisotropic it makes a
great deal of difference just exactly where the axes are; for a fixed
biological direction in the object, a small rotation of the axes
affects the predicted directional growth roughly in proportion to the
difference of the two canonical dilatations. A natural way to construe the discordance of axes, then, is as the discrepancy between
the actions of the two shape changes on a set of directions. Which
directions? Presumably, the principal axes of the derivatives themselves, in proportion each to its own weight. All these considerations are satisfied by the derived matrix $A_1A_2 - A_2A_1$, the commutator of A_1 and A_2. This well-known item is zero whenever the axes of
A_1 and A_2 are perfectly aligned, just as our measure ought to be.
In terms of the parameters β, a_i, b_i suggested above, the commutator
may be computed to be

$$\sin \beta(x,y) \cos \beta(x,y) (a_1(x,y)-b_1(x,y))(a_2(x,y)-b_2(x,y)) \begin{pmatrix} 0 & -1 \\ 1 & 0 \end{pmatrix}$$

and is characterized fully by that scalar multiplier out front.

But $\sin \beta \cos \beta = .5 \sin 2\beta$. In effect, use of the commutator
weights the straight angular discrepancy $|\sin 2\beta|$ by two chances for
it not to matter: A_1 nearly isotropic at (x,y) or A_2 nearly isotropic
at (x,y).

This measure may be standardized in either of two ways. Globally,
for an average discordance suitable for comparing among several analyses, we might simply use the multiplier in a weighted-mean computation:

$$\overline{DC} = \frac{\int\int (a_1 - b_1)(a_2 - b_2)|\sin 2\beta|dx \, dy}{\int\int (a_1 - b_1)(a_2 - b_2) \, dx \, dy}$$

This quantity will vary between 0 and 1. Locally, for comparisons within a single analysis, we might divide the commutator coefficient by its theoretical maximum $a_1 a_2$ to generate the scalar field

$$DC(x,y) = |\sin 2\beta| (1 - b_1/a_1)(1 - b_2/a_2) \ ,$$

again ranging between 0 and 1. It is nearly zero whenever either the axes are nearly in alignment or one of the tensor fields is nearly isotropic. It is near 1 when the angular discrepancy is close to $\pi/4$ and both dilatations are highly anisotropic.

Linear methods. Such a procedure allows the detailed comparison of these tensor representations of growth taken two at a time. It is also meet to have analogues of linear statistical modelling for large collections of tensor fields. It must be possible to add these together, "factor" them, average them, locate mean differences, etc. It ought to be possible to deal with them as dependent variables to be "partitioned" and "explained" in analogues of the analysis of variance. For instance, we should be able to fit a series of powers of the same growth field to a given series of shapes in some optimal way, and to refer to the "fitted" growth and the "unexplained" growth for each transformation. We also might perform a morphometric analysis of covariance, optimally dividing out some sort of allometric growth before fitting a secular trend.

I am currently implementing such a general linear model for systematic spatial samples of tensor fields. The machinery is appropriate for the case of multiple images II_i, $i=1, 2, \ldots, N$, considered as transforms of the same starting image I. Select points P_j, $j=1$, $2, \ldots, J$, in image I. Let A_{ij} be the affine derivative of the transform $I \rightarrow II_i$ at point P_j. The set of A_{ij} for all i is then a tensor-valued variable, $A_{.j}$, while the set of A_{ij} for all j is a J-point sample $A_{i.}$ of the map $I \rightarrow II_i$.

Linear relations among the $A_{.j}$ might be modelled by expressions of the form

$$A_{.J} \sim R_1 A_{.1} + R_2 A_{.2} + \ldots + R_{J-1} A_{.J-1} \ ,$$

meaning

$$A_{iJ} = R_1 A_{i1} + R_2 A_{i2} + \ldots + R_{J-1} A_{iJ-1} + E_i$$

where the R_j are constant 2 x 2 matric coefficients and each E_i is
a tensor-valued "residual." We may <u>fit</u> this model by computing the
R's so that the E's are "smallest" in some sense. To this end we
need scalar quantities to represent the magnitudes of the general
matrices E_i. A good candidate is $||E|| = tr(E^tE)$; this value is
positive-definite and invariant against rotation of image I or the
images II_i separately. If we assemble supermatrices

$$A = \begin{pmatrix} A_{11} & \cdots & A_{1\overline{J-1}} \\ & & \\ & & \\ & & \\ & & \\ A_{N1} & \cdots & A_{N\overline{J-1}} \end{pmatrix},$$

$$R = \begin{pmatrix} R_1 \\ \cdot \\ \cdot \\ \cdot \\ R_{J-1} \end{pmatrix}, \qquad B = \begin{pmatrix} A_{1J} \\ \cdot \\ \cdot \\ \cdot \\ A_{NJ} \end{pmatrix},$$

it can be shown by an analogue to the ordinary normal-equation scheme

(1) that the solution for the R's that yields least mean $||E||$
is $R = (A^tA)^{-1}A^tB$;

(2) that a certain matric "variance" B^tB partitions into an unexplained part, $\sum_i E_i^t E_i$, and an explained part, $\sum_i P_i^t P_i$, where $P_i =$
$R_1 A_{i1} + \ldots + R_{J-1} A_{iJ-1}$ is the ith "predicted value";

(3) that these two variances can be compared for significance
of the regression, or of additional predictors added to the regression, using an approximate Wilks' Λ under a model in which the elements of E are jointly normally distributed.

Applications of this statistical method and its natural generalization to principal components models will concern me in subsequent
publications.

B. <u>Computation</u>

Production of biorthogonal grids using the program described
here, while routine and unproblematic, is far from a mature technique.

For instance, the curving of boundaries ought not generally to be restricted to the class of conic arcs, which disallow inflections (points of zero curvature) and most other curvature profiles. Suitable early generalizations might provide conic splines whose knots are not required to be landmarks, or cubic curves in place of quadratics. Ultimately the program should support boundaries smoothly specified in any manner allowing quick computation of intersections with transversal straight segments, of "inside" and "outside."

Other kinds of information about homology. A more fundamental difficulty in the current implementation is its restriction upon forms of prior information about homology. Ways must be found to evade the null assumptions of the method--linearity of boundary homology with arc-length, devolution of interior information upon points only. An organism, for instance, may have spatial patterns--banding, spacings of cell walls, location of setae--which, though not reliable enough to serve as landmarks strictly comparable from instance to instance, still clearly indicate gradients of local relative growth which ought to supersede the smooth linearity currently imputed inside. Whenever a map is intrinsically unsmooth (unharmonic), as in Fig. VIII-1, my algorithm will not interpolate correctly unless the facts about unsmoothness are input somehow a priori, as by the somewhat inconvenient construction of pseudolandmarks in a dotted line across the form.

For a first essay at generalizing smoothness to resolve these difficulties, I suggest a family I have labelled laminar transforms. Imagine the assignment, upon a region of tissue, of proper curvilinear coordinates (x_b, y_b) whose level curves are just the parallels of a biorthogonal grid. Let f_1 be a monotone function which maps the full range of x_b's onto itself, and f_2 likewise the y_b's. The transform which takes the point of curvilinear coordinate (x_b, y_b) into $(f_1(x_b), f_2(y_b))$, as in Fig. VIII-1, is not smooth, for $\nabla^2 f_1 + \nabla^2 f_2$ will not be zero unless f_1 and f_2 are both linear functions separately. Maps involving such laminations, which usefully model certain axial and meristematic processes, would be estimated correctly by minimum-Laplacian computations only if such a lamination were expressly "partialled out" of the data beforehand. If forms I, II are related by biorthogonal grids G_I, G_{II}, a laminar transform of either form, aligned with its grid G, will leave unaltered the biorthogonal grids of the transform. Then it may be possible to construct an

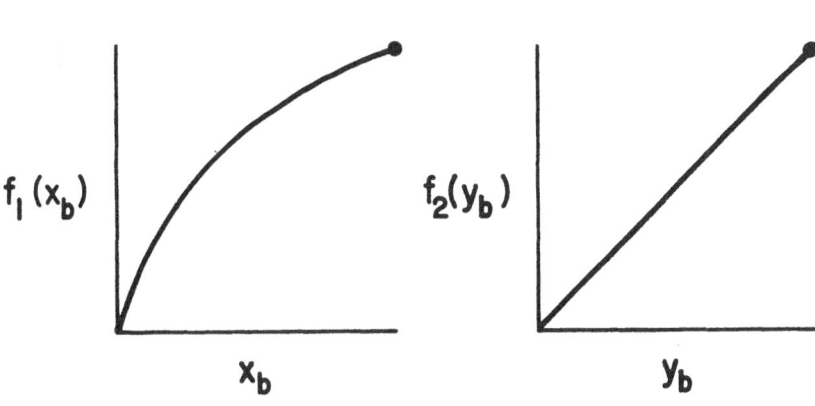

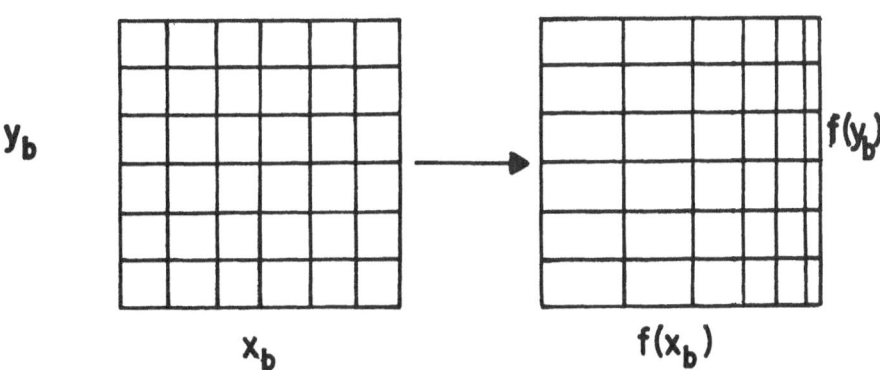

Fig. VIII-1.--Diagram of a laminar transformation.

algorithm which smooths the interpolation and estimates the lamination at the same time.

A further subtlety of real data is the provenance of interior
information in the form of extended arcs homologous as whole structures (this homology, too, perhaps nonlinear in arc-length). For
the comparisons among primate skulls reported in chapter vii, we·
were not able to make use of any data on the path of the cranial
base across the form, except at sella, or of the loci of the orbits
or the zygomatic arches as projected on the midplane. Nor can one
simply declare these internal boundaries equivalent to external ones,

that is, partition the image into sub-regions to be analyzed separately. One could then not enforce differentiability of the resulting
composite map across the sides of what is really the internal boundary; the problem would be remedied by an algorithm for establishing
just such a constraint in otherwise independent analyses.

    <u>Three dimensions</u>. In three dimensions these flexibilities of
specification of homology diversify and proliferate rather as do the
styles of landmarks themselves, described in chapter ii. But the
subtle details of interpolation are less important than structures
for curvilinear coordinates on the surfaces of solid objects: the
extension of the biorthogonal method to closed unbounded manifolds.
The skull, for instance, is an irregular ball, not the inside of a
curve on a picture plane as we have been drawing it. The axes of
the anthropoid engridments in chapter vii are in biological fact
all drawn obliquely upon the curving surfaces of the crania; projected back upon those surfaces, they would no longer be orthogonal.
The computations then apply only to the line drawings which are
their data.

    In extending the biorthogonal method to handle three-dimensional
data, it is not necessary to alter the mathematical argument in any
essentials. In the planar case, the subject of all my examples here,
the boundary by itself is not a rigid structure, and its changes in
length do not determine its changes in shape. It is not enough to
know at what rates it is growing all around, but in which directions
as well. Then the biorthogonal engridment is necessarily in terms of
differential growth throughout the inside of the form. When we add
a dimension we encounter the surfaces of solids, which are curved
two-dimensional manifolds. These forms will in general be more or
less convex. Now closed strictly convex surfaces are mathematically
rigid. If geodesic distances be specified between points in the surface, then the whole structure is fixed up to an equiform transformation. (The demonstration of this is known as the Problem of Weyl:
cf. Stoker, 1969: sec. x.6.) If we specify length changes on a surface, the ensuing solid form change is fully determined, including
all bulging. In short, for three dimensions change in shape is
wholly specified by change in surface distances--this was not true
for outlines in the plane. Hence our grid, too, can be restricted
to the surface itself; we need not construe "pressures from the interior" as in the metaphors of the planar analysis. This is just as
well in view of the great practical difficulty of collecting mor-

phometric data from the insides of living biological forms.

In three dimensions, then, as in two, the Thompson problem is effectively one of transformations on surfaces. Now locally any reasonable surface can be mapped conformally onto the place. (The cartographers discovered this long before the mathematicians. Cf. Kreyszig, 1968: secs. 57, 58.) In such a map, distances may be distorted, but by the same factor in all directions away from any point; then angles are preserved. Two corresponding neighborhoods on two curved surfaces may each be mapped conformally onto planes. The biorthogonal axes through neighborhoods are defined by reference only to angles and so are unchanged by any such combination of maps. Then the very same mathematical theory of coordinate form applies locally: the distorted grids and the singularities we have already seen are just to be pasted upon the curved forms. One need not even execute the conformal mapping in the course of computation. The formulae are set forth entirely in terms of tangent spaces, precisely so that they might generalize verbatim to surfaces in three-dimensional space.

The only difference between the analysis in two dimensions and the analysis in three is that closed surfaces in space have no boundaries. Coordinate curves of a biorthogonal grid extend indefinitely, winding around and around the form in an ergodic net. Furthermore, no coordinate system without singularities can apply to a simple closed surface anyway, and in particular no biorthogonal system could possibly do so. (This is Poincaré's Theorem: cf. do Carmo, 1976: sec. 4.5.) Then we must restrict the analysis to more or less local patches of surface, perhaps hemisphere-sized, set at the discretion of the investigator.

To execute a biorthogonal analysis in three dimensions will be straightforward once we have a one-to-one twice-differentiable correspondence between empirically measured surfaces. Techniques for the computation of such correspondences in useful form are prologue to the biorthogonal method, not properly part of the technique itself. There seems, however, to be no published algorithm to refer to. I have suggested one in section B of chapter iv. Unfortunately, not all shapes of interest are sufficiently "blobby." The task will be made quite difficult by the inclusion of such morphological infelicities as holes, processes, shells, and edges. These complicate the numerical geometry beyond the belief of anyone who has never tried it.

The method of biorthogonal grids provides a symmetric tensor field for surface pairs. But any surface already has a symmetric tensor field all its own, namely, its first fundamental form, its Gauss-Riemann metric tensor. The growth tensor is simply the quotient of the metric tensor from image II by the metric tensor from image I, point for point. We can interpolate or extrapolate shape change, then, by multiplying the metric tensor for image I by a suitable power p of the growth tensor, arriving at a new surface whose relation to image I is that of the pth power of the shape change that resulted in image II. (We can also factor in spatially varying scale changes if we have reason to do so, and we can multiply together two growth processes to create a third.) The problem of constructing the surface that this new product tensor describes is solved in principle by the Riemann Embedding Theorem, which declares that any reasonable metric can be realized in patches by a smooth surface in E^3. The literature provides no hint of a way for constructing such a surface; so another future task I have set is the construction of a program for executing the embedding guaranteed by the theorem.

C. Likely Applications

With any of these improvements, the method of biorthogonal grids clearly can help remedy the presently ungeometrical content of shape and growth analysis. Several applied problems cry out for a sound morphometrics: orthodontia, normal and abnormal embryology, computed tomography, plastic surgery, and all the many special sciences of specific organs, animal and vegetable. I shall discuss three of these possibilities.

Computed tomography. This newest technique of medical imagery has become routine in our larger medical centers only in the last three years. An array of x-ray emitters and detectors rotates completely around a biological object (a human head, for instance), gathering information always in a section (slice) of some fixed thickness along the axis of rotation. Immediately at the conclusion of the scan, a minicomputer, using any of several algorithms, reconstructs a fair picture of the x-ray absorption densities all through the slice. See Hounsfield and Ambrose (1973), Ledley et al. (1974), Brooks and DiChiro (1975), Gordon et al. (1975), and Wood (1976). Insofar as different tissues, or healthy and diseased states of the same tissue, have different absorptivities, this provides good in-

formation about what sorts of tissues lie where in the section. By
a succession of closely spaced slices, furthermore, we can provide
for the same detail along the axis of rotation as transverse to it:
a complete three-dimensional reconstruction of the anatomy. The results are usually displayed slice by slice on a cathode-ray screen,
with different absorptivities coded by grey level or by color. Familiar organs are thereby realized as homogeneous areas of some standard density code. In effect the section is visualized, transformed
into a convincingly real image, the configuration of bones and tissues just as in life. This sort of computation is now proceeding
thousands of times a day and accumulating huge numbers of sections
normal and variously abnormal. Each image is fully digitized (for
that is the way it was computed) and is stored on magnetic tape in
the event that the radiologist wishes to re-visualize it.

What a fabulous data base this could be for statistical studies
of anatomy and its variations, were there a morphometric grammar of
empirical shape! Any instrument will accumulate the images of
hundreds of cerebral ventricles, hundreds of lungs and livers within
their body walls. Surely there is fundamental biomedical information
to be gained from careful study of the variation of these configurations! But, with all the data perfectly represented, tissue for tissue, and readily at hand, there is in the medical, statistical, and
mathematical literature no method whatever for the systematic comparison of the shapes thus encoded, neither in section nor in all
their round fullness. Every month now, indexed under "Tomography,
computed" in Index Medicus, there appear dozens of clinical reports--
ten cases of such-and-such a disease and how their computed sections
indicated this painlessly--but nowhere is there the basic methodology
for shape analysis that would allow us to extract system from the
data without diagnostic preconceptions. We could talk then quantitatively of "the normal lung" or "the normal ventricle" as a statistical range of shapes, configured in a position within the body
cavity likewise having a range. We could follow all manner of bulges
and flexes across individual or collective histories from fetus to
adult, if we only had a metric vocabulary. The method of biorthogonal grids, applied to comparisons of shapes with norms, supplies
just such a vocabulary, flexible and visible.

The radiological profession, accustomed to extracting sense
from indefinite superimposed shadows, has little difficulty scanning computed sections in search of various things, symptoms, which
shouldn't be there. But there should be rich diagnostic information

as well in the shapes and positions of the tissues which should be
there, and for this there are no standards even in two dimensions.
In three dimensions, where only the surgeon or anatomist has formal
training, the challenge is much greater. In the near future we will
be able to image the beating heart in space: "pull it out of the
body," depict it in isolation, by a stereo computer reconstruction,
beating away in full view from any angle. Cf. Glenn et al. (1975),
Adams et al. (1976), Robb et al. (1976). How are we to apprehend
such a splendidly informative phantom? How do we set up descriptors
for such a subtle construct as a periodic change of shape? Intuition and clinical experience both are poor guides when an entirely
new modality emerges: recall the struggle to make sense of the elec-
trocardiogram. The morphometrician must be involved in tomography,
simply to make use of the information that is there. And among the
special tools of his trade will surely be some method of quantifying
shape change comparable to the technique of biorthogonal grids.

Orthodontics. All predictive methods try to unearth some quantity which seems to behave consistently, then extrapolate algebraic
consequences of its continuing to do so. The purpose of any initial
measurement is the construction of ostensible constants for extrapolation. The usefulness of a quantitative prediction, such as the
orthodontist computes for the course of natural or "corrected" growth
in a patient, depends not only upon the accuracy of the original
measures but also upon the validity of the invariants. Naturally it
helps if the "constants" truly have been constant and the clinician
can apply to the problem some more powerful tool than "clinical experience," intuition, or Delphic prophecy. But it must also be the
case that the constants be estimable from the data one is using. If
the initial measures are of a single cephalogram, the necessary constants must be represented there, waiting to be extracted.

In the absence of any analytic technique for the quantification
of facial growth per se, such as has been described in these pages,
all the methods currently available construct constants from shape
data. A single cephalogram is measured, and extrapolation is by way
of unvarying algebraic manipulation. One sometimes hears of "class
III growth," "vertical growers," etc. These terms correspond to the
intuition that there are typical growth patterns associated with
various morphological configurations. Alternatively, one may, without any labelling, adjust the measured values to anticipate expected
change. The values being adjusted may be scalars (distances and

angles) from a tabular analysis like Björk's or Downs's (cf. Merow, 1975), or vectors from a tracing scheme like Ricketts' (1975) or Johnston's (1975). The adjustments may be means from some population of reference (age-, sex-, locale-, race-, or Angle-class-specific), or they may be regression estimates manifesting covariance adjustments themselves for other measures, perhaps mandibular form, often viewed as leading or typing facial growth elsewhere.

All these conventional methods of prediction are based on a single cephalogram, representing the craniofacies at one age only. The growth prediction derives from analysis, statistical or intuitive, of a population pool of completed case histories not including the patient under study. These histories embody growth rates themselves not analyzed. Then the prediction is reliable only insofar as the constants delineating the growth are consistent across the reference population and are appropriate to the specific patient at hand. But these constants have never been measured, even adumbrated, anywhere in the cephalometric literature. There is no way of verifying the necessary assumptions for a particular patient. It is no wonder that the probable error of prediction is so depressingly high (Hirschfeld and Moyers, 1971).

Constants of growth go unexamined because they are geometrically difficult to visualize. All the quantities used in current predictive methods derive from very simple geometry--all can be measured from the cephalogram using ruler, compass, and protractor. This geometric simplicity goes against all the evidence. Actual growth is geometrically quite complex. Enlow's researches (see chapter v) show how general curved form moves through bone, subtly adjusting proportions as it passes. Moss and Salentijn (1970) have identified a logarithmic spiral that aids in the description of mandibular growth, and Ricketts (1972) a circular arc for which he suggests the same usefulness. It is easy to show by example that growth by constant motion upon either of these arcs is not describable by constants in any of the linear cephalometric schemes. No real growth is geometrically straight, nor can it be properly detected by the techniques criticized in chapter iii. Invariant measures are likely buried in nonlinear relationships within a sequence of cephalograms. General models for change are irrevocably curved, and the appropriate mathematics is freed from lines and circles throughout. Instead distances need to be measured along specific curve systems which are determined from the facts of the growth changes themselves.

The patterns of growth the biorthogonal method might extract
from data are biological rules for getting from one form to a subsequent one. For valid prediction we simply have to have them in
quantitative form. When we compute a quantity of these patterns we
may discover that they are the same for large segments of the population, or we may find that they vary in loose association with
features of the single cephalogram. We cannot expect that the patterns are implicit in the single cephalogram. They are expressing
different systems of biological variability, and neither shape nor
shape change can be estimated from the other without systematic error. We need to measure both.

The stubborn refusal of craniofacial growth to cooperate with
cephalometric prediction implies that extrapolations from a single
cephalogram, with any algorithm whatever, are overlooking necessary
information. The same idiosyncrasies which cause a shape to vary
howsoever from the norm will be associated with variation in the
growth tensor leading from that shape onward. It would be sounder
to start from two snapshots of growth--a child at six years of age,
the same child at ten--and to estimate thereby the scale of his particular dilatations and the locus of his lines. I suspect that the
lines are relatively more variable over persons but more stable over
time, that growth spurts may alter the dilatations but will keep
perpendiculars perpendicular. The child's personally computed
growth tensor may then be submitted to whatever statistical manipulations are deemed suitable for extrapolation from age ten to
adulthood. There will result a predicted shape at maturity based on
not only the population norms but also the vagaries of individual
developmental history.

Ordinarily the orthodontist does not have a sequence of cephalograms on which to base an estimate of the growth operator for his
patient. Even so, it is not valid to predict change of shape by adjustment of cephalometric measures only. Instead the clinician must
enter into a two-stage procedure, first estimating the growth operator, then applying this estimate to the cephalogram at hand to predict future form. Such a procedure will at first seem tedious. It
will require a considerable expansion of reference tables and curves.
This orthodontic application of the tensor techniques will not come
quickly, but it clearly needs doing. The clinician would of course
not perform the computations himself; he would send single cephalograms and series to a central processing facility where, as a patient

is followed up, a central data bank is updated with the results of
another empirical dilatation history.

There are several clinical problems in facial growth which could
reasonably be expected to yield to an operator-based model of shape
change. All can be described as direct adjustments of the pattern of
growth irrespective of the original form. Occlusal changes through
time cannot be assumed to be synchronous with craniofacial skeletal
changes. There are also growth "spurts," sudden changes of rates
within the same geometric pattern. In linear formulations, spurts
are seen as abrupt, "inexplicable" change, after the manner of Fig.
III-8(a), (b), or (c). They may instead be sudden changes of rate
of growth along invariant curves, a problem rather easier to handle.
Finally, the student of facial growth is not yet able to deal with
the impact of orthodontic treatment upon growth. Knowledge of the
natural patterns of growth must be applied to treatment protocols,
to incorporate into predictions the pattern changes caused by treatment.

Developmental biology. My analysis of the hominoid skull series
hints tantalizingly at a hypothesis which cannot yet be tested. Each
skull of the series is the end-product of its own ontogenetic shape
change. It is possible that the axes describing the grosser changes
over evolutionary time express an intrinsic natural variability
operating at the ontogenetic level. Then the shape changes of
evolution might be the result of selection for particular gradients
in a system of form variation which has rather few degrees of
freedom--surely more than the two or three I have located in my crude
cranial figures, but fewer than, say, a dozen. Evolution by adjustment of the stuff of allometry--relative rates and timings--in an
otherwise unchanging developmental system is not a new concept: for
a review, see Gould (1977). The biorthogonal method extends its geometric power from scalars to tensors, and may make possible an enumeration of features of interest before the onset of numerical measurement. In this context the method expresses geometrically the
coordinations intrinsic to a whole collection of ontogenies, and
thus specifies loci along which to look for biomechanical causes,
specific abnormalities, and the like.

This speculation relates to a larger subject, spatial positional
information in development--the provenance of spatial coordinates in
one form or another to cells whose differentiation seems to depend
on position. Cf. Wolpert (1969, 1971), Cooke (1975). Several re-

searchers currently are studying the actual coordinate systems upon
which positional information is borne throughout the interiors of
extended structures. They often conclude that the general shape of
the organ is a crucial determinant of its pattern of information and
differentiation. For instance, Kauffmann, Shymko, and Trabert (1978)
model the formation of compartments in the wing disk of _Drosophila_
according to the formalism of spatial eigenfunctions for a diffusion-
reaction system; Schwartz (1977) models the mapping between retina
and optic tectum in goldfish, long studied by Sperry, Gaze, and
others, by way of a conformal (angle-preserving) distortion which,
like the interpolations invoked in this essay, is optimally smooth
for its boundary conditions. In both these proposals, the boundaries (coincidentally taken as elliptical) do not bear any information themselves, but only bound: the sole coordinates they supply
are their own spatial locations. The formalism of biorthogonal
grids provides a quite different alternative for the provenance of
information to differentiating systems extended in space. The directional growth-gradients reconstructed by the biorthogonal method
correspond to the trace of two or more local coordinate systems which
might supply spatial information in precisely this sense. They may
serve as model for actual morphogenetic gradients, just as I located
a field concentric about the tail in the reanalysis of _Diodon_, Fig.
VII-1. It would be of interest to run numerical experiments in
which growth-gradients are plotted which fall off exponentially (as
in diffusive processes) with distance from some small number of centers. Varying the positions and the strengths of these centers would
provide a family of biorthogonal coordinate systems in which certain
conventional patterns, such as fingerprints, might well be recognized.

In the developmental biology of the lower forms, explanatory
theories of normal budding and of various experimentally produced
anomalies often invoke the formal device of inferred polarities,
directional gradients of some postulated morphogenetic regulator.
Though these are usually discussed in systems of a single axis, the
data are of creatures with the usual three dimensions of extent.
The formalism of biorthogonal grids provides a method for aggregat-
ing the effects of multiple growth-centers distributed in a plane or
a curving manifold, and may thus explain geometrically certain situations in which polarities are competing, such as experiments in
grafting and severing. Cf. Berrill (1961: ch. 3 and part ii).

The early vertebrate embryo is the most intensively studied system in developmental biology. The literature (cf. Kühn, 1971) contains a hundred years of qualitative description of normal development together with a certain amount of mainly univariate quantification. The shape changes of the early stages (blastulation, gastrulation) are especially well-described, and in particular, "fate maps" exist tracing typical cells or regions through the changes stage by stage. There is just now beginning to be serious biomathematical modelling of these processes: cf. Thom (1975: ch. 9), Zeeman (1974), or Jacobson and Gordon (1976). Such modelling could be performed instead by computation of a suitable biorthogonal system for whole thick sheets of cells. The adjustment of cell shape, in which cylinders may change their aspect ratio or be deformed into truncated cones, "bottle cells," or worse, does not necessarily drive this process, as in the Jacobson-Gordon simulation I reviewed in chapter v, but may instead be bound up in a system for differential shape change at top and bottom of a tissue one cell thick. The biorthogonal method can easily extend to such bending of sheets of cells and provide rigorous geometrical descriptions of all manner of curving tissues: early, the blastula and gastrula; later, the neural tube, the eye cups, regenerating blastemae, and all manner of other tissues whose curvature changes with growth. Gordon calls this phenomenon "morphodynamics" and models it by a creeping viscous fluid. I believe that the biorthogonal method of analysis by growth tensors will provide a more useful level of quantitative spatial structure.

CHAPTER NINE. ENVOI

The human eye is wired for <u>Gestalt</u>, for recognition rather than
quantification. It is notoriously bad at apprehending population
variance. A catalog or key setting forth a variety of typical forms
for a phenomenon, such as that for leaf shape in Fig. IX-1, does
not specify in any way how one might describe forms within a family
(e.g. the growing leaf, its shape constantly changing) or assign
particular specimens, always more or less idiosyncratic, to particular types. Such a classification should rather arise from a thoroughgoing quantification along the various lines I have pursued in this
essay. The study of all form is the study of comparative or growing
form, and the necessary quantifications are all based in the geometry
of ordinary planes and space.

The various types of machinery I have proposed in this essay--
curvature functions, skeletons, biorthogonal grids and the growth
tensor--are but samples of what the geometer's untapped store of
constructs and representations promises for morphometrics. His notions match biomedical problems of form and form change much more
fruitfully than do the current analytics based on Euclidean measures
and linear statistics. It is not difficult to assimilate the necessary background in post-Cartesian geometry; it could be learned
from survey courses in less time than the better biomedical graduate
students now devote to advanced calculus and linear algebra for multivariate statistical analysis. And modern geometry provides the
intuitive imagery necessary to link measurement and analysis: specific formalisms for what we see and the way we record it. Quantification of Thompson's method of transformation grids has been
waiting all century for an infusion of geometric insight, and other
problems, I am sure, are as ripe for fresh methods towards their
regularization. Morphometrics can be made over entirely using applied geometry: made over into an intellectual tool sensitive to
natural variation in the systems studied yet compatible with the
quantitative spirit. Puzzles of morphometrics have captivated biologists for as long as there has been geometry. By now, with the
sixtieth anniversary of <u>On Growth and Form</u> upon us, it is high time
to bring the two fields into fruitful juxtaposition. I hope that
this essay, record of my first few discoveries, hastens the advent of
a fully geometrical science of form.

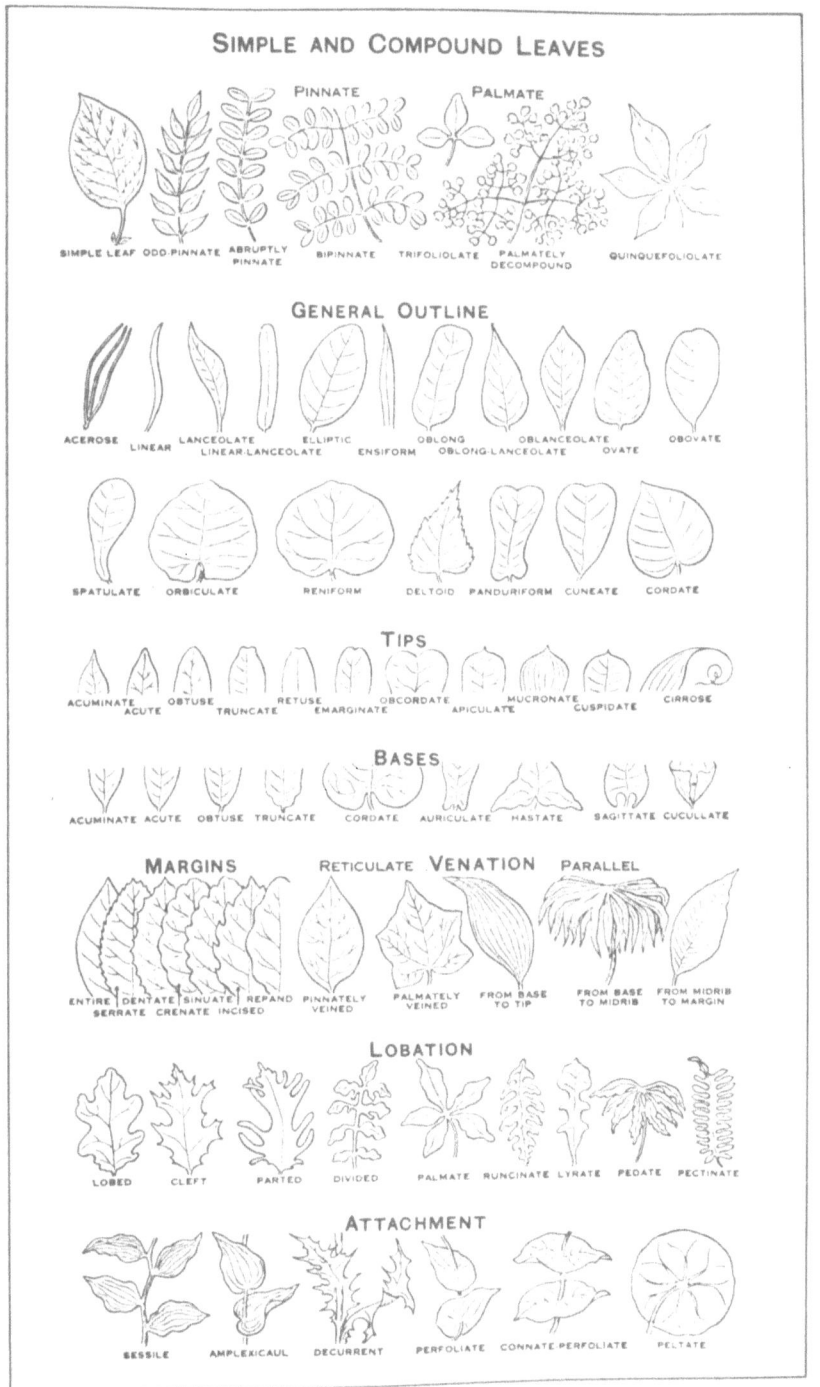

Fig. IX-1.--The shapes of leaves. By permission. From Webster's New International Dictionary, second edition, © 1959 By G. & C. Merriam Co., publishers of the Merriam-Webster dictionaries.

LITERATURE CITED

Abbie, A. A. (1963), "The cranial centre", Z. Morph. Anthrop. 53:6.

Abramowitz, M. A., and I. C. Stegun (1964), Handbook of Mathematical
    Functions (Washington, D.C.: National Bureau of Standards).

Adams, D. F., et al. (1976), "Differing attenuation coefficients of
    normal and infarcted myocardium", Science 192:467.

Agin, G. J., and T. O. Binford (1976), "Computer description of
    curved objects", I.E.E.E. Trans. Comp. C-25:439.

Ahlberg, J. H., E. N. Nilson, and J. L. Walsh (1967), Theory of
    Splines and their Applications (New York: Academic Press).

Akima, H. (1974), "A method of bivariate interpolation and smooth
    surface fitting based on local procedures", Comm. Assoc. Comp.
    Mach. 17:18.

Albano, A. (1974), "Representation of digitized contours in terms of
    conic arcs and straight-line segments", Comp. Graphics Image
    Proc. 3:23.

Appleby, R. M., and G. L. Jones (1976), "The analogue video reshaper--
    a new tool for palaeontologists", Paleontology 19:565.

Attneave, F., and M. D. Arnoult (1956), "The quantitative study of
    shape and pattern perception", Psych. Bull. 53:452.

Avery, G. S., Jr. (1933), "Structure and development of the tobacco
    leaf", Am. J. Bot. 20:565.

Bachi, R. (1962), "Standard distance measures and related methods
    for spatial analysis", Regional Science Assoc, Papers, X, Zurich
    Cong.

Baer, M. J., and S. K. Nanda (1976), "A commentary on the growth and
    form of the cranial base", in Development of the Basicranium,
    ed. J. F. Bosma (Bethesda, Md.: U.S. Department of H.E.W.).

Barnhill, R. E. (1977), "Representation and approximation of surfaces", in Mathematical Software III, ed. J. R. Rice (New York: Academic Press.)

Bennett, J. R., and J. S. Macdonald (1975), "On the measurement of curvature in a quantized environment", I.E.E.E. Trans. Comp. C-24:803.

Berrill, N. J. (1961), Growth, Development, and Pattern (San Francisco: W. H. Freeman and Co.).

Biegert, J. (1957), "Der Formwandel des Primatenschädels", Gegenbaurs Morph. Jahrb. 98:77.

Biegert, J. (1963), "The evaluation of the characteristics of the skull, hands, and feet for primate taxonomy", in Classification and Human Evolution, ed. S. L. Washburn (Chicago: Aldine Publishing Co.).

Biggerstaff, R. H. (1972), "Three variations in dental arch form estimated by a quadratic equation", J. Dent. Res. 51:1509.

Bingham, C. (1974), "An antipodally symmetric distribution on the sphere", Ann. Stat. 2:1201.

Björk, A. (1969), "Prediction of mandibular growth rotation", Am. J. Orthod. 55:585.

Björk, A., and V. Skieller (1972), "Facial development and tooth eruption", Am. J. Orthod. 62:339.

Blackith, R. E., R. G. Davies, and E. A. Moy (1963), "A biometric analysis of development in Dysdericus fasciatus sign", Growth 27:317.

Blackith, R. E., and R. A. Reyment (1971), Multivariate Morphometrics (London: Academic Press).

Blum, H. (1973), "Biological shape and visual science", J. Theor. Bio. 38:205.

Blum, H., and R. N. Nagel (1977), "Shape description using weighted symmetric axis features", Proc. I.E.E.E. Comp. Soc. Conf. Pattern Recognition and Image Processing [77CH1208-9C].

Bookstein, F. L. (1977), "Orthogenesis of the hominids: an exploration using biorthogonal grids", Science 197:901.

Briggs, I. A. (1974), "Machine contouring using minimum curvature", Geophysics 39:48.

Brooks, R. A., and G. DiChiro (1975), "Theory of image reconstruction in computed tomography," Radiology 117:561.

Brown, D. R., and D. H. Owen (1967), "The metrics of visual form: methodological dyspepsia", Psych. Bull. 68:243.

Brown, V., and R. G. Davies (1972), "Allometric growth in two species of Ectobius", J. Zool. Lond. 166:97.

Bunge, W. (1962), "Theoretical geography", Lund Studies in Geography, Series C, General and Mathematical Geography 1.

Burch, J. (1975), Freshwater Sphaeriacean Clams of North America (Hamburg, Michigan: Malacological Publications).

Calabi, L., and W. E. Hartnett (1968), "Shape recognition, prairie fires, convex deficiencies, and skeletons", Am. Math. Monthly 75:335.

do Carmo, M. P. (1976), Differential Geometry of Curves and Surfaces (Englewood Cliffs, N.J.: Prentice-Hall).

Chow, C. K., and T. Kaneko (1972), "Boundary detection of radiographic images by a threshold method", Proc. I.F.I.P. Cong. 1971 2:1530.

Clark, W. A. V., and G. L. Gaile (1973), "The analysis and recognition of shapes", Geografiska Annaler 55:153.

Colbert, E. H. (1935), "Siwalik mammals in the American Museum of Natural History", Trans. Am. Philos. Soc. 26:1.

Collatz, L. (1960), The Numerical Treatment of Differential Equations, 3rd ed. (Berlin: Springer-Verlag).

Cooke, J. (1975), "The emergence and regulation of spatial organization in early animal development", Ann. Rev. Biophys. Bioeng. 4:185.

Cooper, D. R., and N. Yalabik (1976), "On the computational cost of approximating and recognizing noise-perturbed straight lines and quadratic arcs in the plane", I.E.E.E. Trans. Comp. C-25:1020.

Corruccini, R. (1975), "Multivariate analysis in biological anthropology: some considerations", J. Human Evol. 4:1.

de Coster, L. (1939), "The network method of orthodontic diagnosis", Angle Orthod. 9:3.

Coxeter, H. S. M. (1961), Introduction to Geometry (New York: John Wiley and Sons).

Delattre, A., and R. Fenart (1960), L'Hominisation du Crâne (Paris: Editions du C. N. R. S.).

Dmoch, R. (1975-6), "Beiträge zum Formenwandel des Primatencraniums mit Bemerkungen zu den sagittalen Knickungsverhältnissen", Gegenbaurs Morph. Jahrb. 121:450, 521, 555, 122:1.

Duda, R. O., and P. E. Hart (1973), Pattern Classification and Scene Analysis (New York: John Wiley and Sons).

Eccles, M. J., M. P. C. McQueen, and D. Rosen (1977), "Analysis of the digitized boundaries of planar objects", Pattern Recognition 9:31.

Elliott, D. (1970), "Determination of finite strain and initial shape from deformed elliptical objects", Geol. Soc. Amer. Bull. 81:2221.

Enlow, D. H. (1968), The Human Face (New York: Harper and Row).

Enlow, D. H. (1975), Handbook of Facial Growth (Philadelphia: W. B. Saunders Co.).

Erickson, R. O. (1966), "Relative elemental rates and anisotropy of growth in area", J. Exp. Bot. 17:390.

Forrest, A. R. (1972), "On Coons and other methods for the representation of curved surfaces", Comp. Graphics Image Proc. 1:341.

Freeman, H. (1974), "Computer processing of line-drawing images", Comp. Surv. 6:57.

Freeman, H., and L. S. Davis (1977), "A corner-finding algorithm for chain-coded curves", I.E.E.E. Trans. Comp. C-26:297.

Fuchs, H., Z. M. Kedem, and S. P. Uselton (1977), "Optimal surface reconstruction from planar contours", Comm. Assoc. Comp. Mach. 20:693.

Glenn, W. V., et al. (1975), "Image generation and display techniques for CT scan data--thin transverse and reconstructed coronal and sagittal planes", Invest. Radiol. 10:403.

Gnanadesikan, R. (1977), Methods for Statistical Data Analysis of Multivariate Observations (New York: John Wiley and Sons).

Golub, G. H., and C. Reinsch (1970), "Singular value decomposition and least squares solutions", Numer. Math. 14:403; reprinted in Wilkinson and Reinsch (1971).

Gordon, R., G. T. Herman, and S. A. Johnson (1975), "Image reconstruction from projections", Sci. Amer. 233(4):56.

Gould, S. J. (1966), "Allometry and size in ontogeny and phylogeny", Biol. Rev. 41:587.

Gould, S. J. (1971), "D'Arcy Thompson and the science of form", New Lit. Hist. 2:229.

Gould, S. J. (1977), Ontogeny and Phylogeny (Cambridge: Harvard University Press).

Gould, S. J., and R. F. Johnston (1972), "Geographic variation", Ann. Rev. Ecol. Syst. 3:457.

Green, P. B. (1965), "Pathways of cellular morphogenesis", J. Cell Biol. 27:343.

Green, P. B. (1969), "Cell morphogenesis", Ann. Rev. Plant Physiol. 20:365.

Green, P. B. (1976), "Growth and cell pattern formation on an axis", Botan. Gazette 137:187.

Gysel, C. (1972), "Genealogie de la cephalometrie", Rev. Belge Mèd. Dent. 27:257.

Helmuth, H. (1970), "Ueber den Bau des menschlichen Schädels bei künstlicher Deformation", Z. Morph. Anthrop. 62:30.

Herron, R. E. (1972), "Biostereometric measurement of body form", Yearbook of Physical Anthropology 16:80.

Hershkovitz, P. (1977), The New World Primates, v. 1 (Chicago: University of Chicago Press).

Hilbert, D., and S. Cohn-Vossen (1952), Geometry and the Imagination (New York: Chelsea Publishing Co.).

Hilditch, C. J. (1969), "Linear skeletons from square cupboards", in Machine Intelligence 4, ed. B. Meltzer and D. Michie (New York: American Elsevier Publishing Co.).

Hilditch, C. J., and D. Rutovitz (1969), "Chromosome recognition", Ann. N.Y. Acad. Sci. 157:339.

Hirschfeld, W. J., and R. E. Moyers (1971), "Prediction of cranio-facial growth: the state of the art", Am. J. Orthod. 60:435.

Hirsinger, V. (1976), "Numerical strain analysis using polar coordinate transformation", Math. Geol. 8:183.

Hofer, H. (1965), "Die morphologische Analyse des Schädels des Menschen", in Menschlichen Abstammungslehre, ed. G. Heberer (Stuttgart: Gustav Fischer Verlag).

Hopf, H. (1955), Selected Topics in Differential Geometry in the Large, lecture notes by T. S. Klotz (New York University).

Hounsfield, G. N., and J. Ambrose (1973), "Computerized transverse axial scanning", Br. J. Radiol. 46:1016.

Howells, W. W. (1973), Cranial Variation in Man (Cambridge, Mass.: Peabody Museum of Archaeology and Ethnology).

Hursh, T. M. (1976), "The study of cranial form", in The Measures of Man, ed. E. Giles and J. Friedlaender (Cambridge: Harvard University Press).

Hutchinson, G. E. (1948), "In memoriam, D'Arcy Wentworth Thompson", Am. Sci. 36:577.

Huxley, J. S. (1932), Problems of Relative Growth (London: Methuen and Co.).

Izard, G. (1950), Orthodontie, 3rd edition (Paris: Masson et Cie.)

Jacobson, A. G., and R. Gordon (1976), "Changes in the shape of the developing vertebrate nervous system analyzed experimentally, mathematically, and by computer simulation", J. Exp. Zool. 197:191.

Jacobson, N. (1953), Lectures in Abstract Algebra (Princeton, N.J.: D. Van Nostrand Co.).

Jardine, N. (1969), "The observational and theoretical components of homology: a study based on the morphology of the dermal skull-roofs of rhipistidian fishes", Biol. J. Linn. Soc. 1: 327.

Jardine, N., and C. J. Jardine (1967), "Numerical homology", Nature 216:301.

Johnston, L. E. (1975), "A simplified approach to prediction", Am. J. Orthod. 67:253.

Kauffmann, S. A., R. M. Shymko, and K. Trabert (1978), "Control of sequential compartment formation in Drosophila", Science 199: 259.

Kingdon, J. (1971), East African Mammals (New York: Academic Press).

Klatt, B. (1949), "Die theoretische Biologie und die Problematik der Schädelform", Biol. Gen. 19:51.

Klein, F. (1927), Vorlesungen über Nicht-Euklidische Geometrie (New York: Chelsea Publishing Co.).

Kowalski, C. J. (1972), "A commentary on the use of multivariate statistical methods in anthropometric research", Am. J. Phys. Anth. 36:119.

Kreyszig, E. (1968), Introduction to Differential Geometry and Riemannian Geometry (Toronto: University of Toronto Press).

Krogman, W. M., and V. Sassouni (1957), Syllabus in Roentgenographic Cephalometry (Philadelphia: Philadelphia Center for Research in Child Growth).

Kühn, A. (1971), Lectures in Developmental Physiology (New York: Springer-Verlag).

Kummer, B. (1952), "Untersuchungen über eine morphologische Reihe in der Phylogenie des Menschenschädels", Homo 3:109.

Kummer, B. (1953), Untersuchungen über die Entwicklung der Schädel- form des Menschen und einiger Anthropoïden [Abh. exakt. Biol. 3] (Berlin: Borntraeger).

Lanczos, C. (1970), Space Through the Ages (London: Academic Press).

Ledley, R. S. (1972), "Analysis of cells", I.E.E.E. Trans. Comp. C-21:740.

Ledley, R. S., et al. (1974), "Computerized transaxial x-ray tomography of the human body", Science 186:207.

Lestrel, P. E. (1974), "Some problems in the assessment of morphological size and shape differences", Yearbook of Physical Anthropology 18:140.

Levi, G., and U. Montanari (1970), "A grey-weighted skeleton", Inf. Control 17:62.

Lull, R. S., and S. W. Gray (1949), "Growth patterns in the Ceratopsia", Am. J. Sci. 247:492.

Mandelbrot, B. B. (1977), Fractals: Form, Chance, and Dimension (San Francisco: W. H. Freeman and Co.).

Marey, E. J. (1895), Movement (London: William Heinemann).

Medawar, P. B. (1945), "Size, shape, and age", in Essays on Growth and Form, ed. W. E. le Gros Clark and P. B. Medawar (Oxford: Clarendon Press).

Medawar, P. B. (1958), "D'Arcy Thompson and Growth and Form", in R. D'Arcy Thompson, D'Arcy Wentworth Thompson: the Scholar-Naturalist (London: Oxford University Press).

Meltzer, B., N. H. Searle, and R. Brown (1967), "Numerical specification of biological form", Nature 216:32.

Merow, W. W. (1975), "Cephalometrics", in Enlow (1975).

Mikhail, E. M. (1976), Observations and Least Squares (New York: IEP-Dun-Donnelly).

Montanari, U. (1968), "A method for obtaining skeletons using quasi-Euclidean distance", J. Assoc. Comput. Mach. 15:608.

Montanari, U. (1969), "Continuous skeletons from digitized images", J. Assoc. Comput. Mach. 16:534.

Moore, D. J. H. (1974), "On the medial axis function for visual patterns", I.E.E.E. Trans. Syst., Man, Cyber. SMC-4:396.

Moore, W. J., and C. L. B. Lavelle (1974), Growth of the Facial Skeleton in the Hominoidea (London: Academic Press).

Moorrees, C. F. A., and L. Lebret (1962), "The mesh diagram and cephalometrics", Angle Orthod. 32:214.

Moorrees, C. F. A., et al. (1975), "The computerized mesh diagram analysis", in Transactions of the Third International Orthodontic Congress (St. Louis: C. V. Mosby Co.).

Mosimann, J. E. (1970), "Size allometry: size and shape variables with characterizations of the lognormal and generalized gamma distributions", J. Am. Stat. Assoc. 65:930.

Mosimann, J. E. (1975), "Statistical problems of size and shape", in Statistical Distributions in Scientific Work, v. 2, ed. G. Patil et al. (Dordrecht, Holland: D. Reidel Publishing Co.).

Moss, M. L. (1973), "A functional cranial analysis of primate craniofacial growth", in Craniofacial Biology of Primates, ed. M. R. Zingeser, which is v. 3 of Symposia of the Fourth International Congress of Primatology, ed. W. Montagna (Basel: S. Karger).

Moss, M. L., and L. Salentijn (1970), "The logarithmic growth of the human mandible", Acta anat. 77:341.

Needham, A. E. (1937), "On relative growth in Asellus aquaticus", Proc. Zool. Soc. Lond. A 107:289.

Needham, A. E. (1943), "On relative growth in Asellus aquaticus. II", Proc. Zool. Soc. Lond. A 113:44.

Needham, A. E. (1950), "The form-transformation of the abdomen of the female pea-crab Pinnotheres pisum Leach", Proc. Roy. Soc. Lond. B 137:115.

Needham, A. E. (1964), The Growth Process in Animals (Princeton, N.J.: D. Van Nostrand Co.).

Olson, E. C. (1975), "Permo-carboniferous paleoecology and morpho-typic series", Amer. Zool. 15:371.

Oxnard, C. E. (1973), Form and Pattern in Human Evolution (Chicago: University of Chicago Press).

Paton, K. A. (1970a), "Conic sections in chromosome analysis", Pattern Recognition 2:39.

Paton, K. A. (1970b), "Conic sections in automatic chromosome analysis", in Machine Intelligence 5, ed. B. Meltzer and D. Michie (New York: American Elsevier Publishing Co.).

Pavlidis, T., and S. L. Horowitz (1974), "Segmentation of plane curves", I.E.E.E. Trans. Comp. C-23:860.

Pearson, K. (1901), "On lines and planes of closest fit to systems of points in space", Philos. Mag. ser. 6 2:559.

Pedoe, D. (1970), A Course of Geometry for Colleges and Universities (Cambridge: The University Press).

Persoon, E., and K.-S. Fu (1977), "Shape discrimination using Fourier descriptors", I.E.E.E. Trans. Syst. Man Cyber. SMC-7: 170.

Peskin, C. S. (1975), Mathematical Aspects of Heart Physiology (New York: Courant Institute of Mathematical Sciences).

Pickert, G., R. Stendor, and M. Hellwich (1974), "From projective to Euclidean geometry", in Fundamentals of Mathematics, v. 2, ed. H. Behnke et al. (Cambridge, Mass.: The M.I.T. Press).

Poirier, D. J. (1973), "Piecewise regression using cubic splines", J. Am. Stat. Assoc. 68:515.

Prewitt, J. M. S. (1972), "Parametric and nonparametric recognition by computer: an application to leucocyte image processing", in <u>Advances in Computers 12</u>, ed. M. Rubinoff (New York: Academic Press).

Proskurowski, W., and O. Widlund (1975), "On the numerical solution of Helmholtz's equation by a capacitance matrix method", ERDA Research and Development Report C00-3077-99, Courant Institute, New York University.

Proskurowski, W., and O. Widlund (1976), "On the numerical solution of Helmholtz's equation by a capacitance matrix method," <u>Math. of Comput.</u> 30:433.

Rao, C. R. (1973), <u>Linear Statistical Inference and its Applications</u>, 2nd ed. (New York: John Wiley and Sons).

Raup, D. M. (1961), "The geometry of coiling in gastropods", <u>Proc. Nat. Acad. Sci.</u> 47:602.

Raup, D. M. (1966), "Geometric analysis of shell coiling", <u>J. Paleontol.</u> 40:1178.

Rensch, B. (1960), <u>Evolution above the Species Level</u> (New York: Columbia University Press).

Richards, O. W. (1955), "D'Arcy W. Thompson's mathematical transformation and the analysis of growth", <u>Ann. N.Y. Acad. Sci.</u> 63:456.

Richards, O. W., and A. J. Kavanagh (1943), "The analysis of the relative growth-gradients and changing form of growing organisms: illustrated by the tobacco leaf", <u>Am. Nat.</u> 77:385.

Richards, O. W., and A. J. Kavanagh (1945), "The analysis of growing form", in <u>Essays in Growth and Form</u>, ed. W. E. le Gros Clark and P. B. Medawar (Oxford: Clarendon Press).

Richards, O. W., and G. A. Riley (1937), "The growth of amphibian larvae illustrated by transformed coordinates", J. Exp. Zool. 77:159.

Richardus, P., and R. K. Adler (1972), Map Projections (New York: American Elsevier Publishing Co.).

Ricketts, R. M. (1972), "A principle of arcial growth of the mandible", Angle Orthod. 42:368.

Ricketts, R. M. (1975), "The application of computers to orthodontics--diagnosis, prognosis, and treatment planning", in Transactions of the Third International Orthodontic Congress (St. Louis: C. V. Mosby Co.).

Riolo, M. L., et al. (1974), An Atlas of Craniofacial Growth (Ann Arbor, Mich.: Center for Human Growth and Development).

da Riva Ricci, D., and B. Kendrick (1972), "Computer modelling of hyphal tip growth in fungi", Can. J. Bot. 50:2455.

Robb, R. A., et al. (1976), "Quantitative imaging of dynamic structure and function of the heart, lungs, and circulation by computerized reconstruction and subtraction techniques", in Proc. 3rd Annual Conf. Comp. Graphics, Interactive Tech., Image Proc.--SIGGRAPH '76, which is Computer Graphics 10(2).

Rogers, D. F., and J. A. Adams (1976), Mathematical Elements for Computer Graphics (New York: McGraw-Hill Book Co.).

Rosenfeld, A., ed. (1976), Digital Picture Analysis (New York: Springer-Verlag).

Rosenfeld, A., and E. Johnston (1973), "Angle detection of digital curves", I.E.E.E. Trans. Comp. C-22:875.

Rosenfeld, A., and A. C. Kak (1976), Digital Picture Processing (New York: Academic Press).

Rosenfeld, A., and J. L. Pfaltz (1966), "Sequential operations in digital picture processing", J. Assoc. Comp. Mach. 13:471.

Rutovitz, D. (1970), "Centromere finding: some shape descriptors for small chromosome outlines", in Machine Intelligence 5, ed. B. Meltzer and D. Michie (New York: American Elsevier Publishing Co.).

Salamon, P., A. List, Jr., and P. S. Grenetz (1973), "Mathematical analysis of plant growth", Plant Physiol. 51:635.

Salzmann, J. A., ed. (1961), Roentgenographic Cephalometrics (Philadelphia: J. P. Lippincott Co.).

Sampson, P. D. (1978), dissertation, University of Michigan, forthcoming.

Sanderson, D. J. (1977), "The algebraic evaluation of two-dimensional finite strain rosettes", Math. Geol. 9:483.

Schüepp, O. (1952), "Wachstum und Zellanordnung im Sprossgipfel", Ber. Schweiz. Bot. Ges. 62:592.

Schüepp, O. (1966), Meristeme (Basel: Birkhauser Verlag).

Schultz, A. H. (1950), "The physical distinctions of man", Proc. Amer. Philos. Soc. 94:428.

Schultz, A. H. (1955), "The position of the occipital condyles and of the face relative to the skull base in primates", Am. J. Phys. Anth. n.s. 13:97.

Schumaker, L. L. (1976), "Fitting surfaces to scattered data", in Approximation Theory II, ed. G. G. Lorentz et al. (New York: Academic Press).

Schwartz, E. L. (1977), "The development of specific visual connections in the monkey and the goldfish: outline of a geometric theory of receptotopic structure", J. Theor. Biol. 69:655.

Schwerdtfeger, H. (1962), Geometry of Complex Numbers (Toronto: University of Toronto Press).

Scott, J. H. (1958), "The cranial base", Amer. J. Phys. Anth. n.s.
    16:319.

Scott, J. H. (1963), "Factors determining skull form in primates",
    Symp. Zool. Soc. Lond. 10:127.

Searle, N. H. (1969), "Shape analysis by way of Walsh functions",
    in Machine Intelligence 5, ed. B. Meltzer and D. Michie
    (New York: American Elsevier Publishing Co.)

Shapiro, B., and L. Lipkin (1977), "The circle transform, an artic-
    ulable shape descriptor", Comp. Biomed. Res. 10:511.

Shiells, K. A. G. (1965), "Growth of a productid shell and its im-
    plication on a method of statistical correlation", Nature
    205:878.

Sneath, P. H. A. (1967), "Trend-surface analysis of transformation
    grids", J. Zool. Lond. 151:65.

Sokal, R. R., and P. H. A. Sneath (1963), Principles of Numerical
    Taxonomy (San Francisco: W. H. Freeman and Co.).

de Souza, P. V., and P. Houghton (1977), "Computer location of medial
    axes", Comp. Biomed. Res. 10:333.

Spanier, E. H. (1966), Algebraic Topology (New York: McGraw-Hill
    Book Co.).

Spivak, M. (1970), Differential Geometry (Boston: Publish or Perish,
    Inc.).

Sprent, P. (1972), "The mathematics of size and shape", Biometrics
    28:23.

Starck, D. (1974), "Die Stellung der Hominiden in Rahmen der
    Säugetiere", in Die Evolution der Organismen, ed. G. Heberer,
    v. 3 (Stuttgart: Gustav Fischer Verlag).

Starck, D., and B. Kummer (1962), "Zur Ontogenese des Primaten-schädels", Anthrop. Anz. 25:204.

Steinhaus, H. (1954), "Length, shape, and area", Colloq. Math. 3:1.

Stoker, J. J. (1969), Differential Geometry (New York: John Wiley and Sons).

Storer, T. (1951), General Zoology (New York: McGraw-Hill Book Co.).

Tan, B. K. (1973), "Determination of strain ellipses from deformed ammonoids", Techtonophysics 16:89.

Teissier, G. (1960), "Relative growth", in The Physiology of Crustacea, ed. T. H. Waterman (New York: Academic Press).

Thom, R. (1975), Structural Stability and Morphogenesis (Reading, Mass.: W. A. Benjamin, Inc.).

Thomas, D. T. (1976), "Pseudospline interpolation for space curves", Math. of Comput. 30:58.

Thompson, D'A. W. (1961), On Growth and Form (orig. 1917, 1942), ed. J. T. Bonner (Cambridge: The University Press).

Tissot, M. A. (1881), Memoires sur les Representations des Surfaces (Paris: Gauthier et Cie.).

Tobler, W. R. (1977), "Bidimensional regression", unpublished ms.

Tobler, W. R. (1978), "The comparison of plane forms", Geographical Analysis, in press.

Verheyen, W. N. (1962), "Contribution à la craniologie comparée des primates", Musée Royal de l'Afrique Central - Tervuren, Belgium - Annales - in 8vo - Sci. Zool., no, 105.

Vogel, C. (1968), "The phylogenetical evaluation of some characters and some morphological trends in the evolution of the skull in catarrhine primates", in Taxonomy and Phylogeny of Old World Primates with References to the Origin of Man, ed. B. Chiarelli (Turin: Rosenberg and Sellier).

Walker, G. F., and C. J. Kowalski (1971), "A two-dimensional coordinate model for the quantification, description, analysis, prediction, and simulation of craniofacial growth", Growth 35:191.

Walker, G. F., and C. J. Kowalski (1972), "A new approach to the analysis of craniofacial morphology and growth", Am. J. Orthod. 61:221.

Webster's New International Dictionary, 2nd edition (1934).

Wechsler, H., and J. Sklansky (1977), "Finding the rib cage in chest radiographs", Pattern Recognition 9:21.

Weidenreich, F. (1941), "The brain and its role in the phylogenetic transformation of the human skull", Trans. Am. Philos. Soc. 35:321.

Widrow, B. (1973), "The rubber mask technique. I. Pattern measurement and analysis", Pattern Recognition 5:175.

Wilkinson, J. H., and C. Reinsch (1971), Linear Algebra (New York: Springer-Verlag).

Wold, S. (1974), "Spline functions in data analysis", Technometrics 16:1.

Wolpert, L. (1969), "Positional information and the spatial pattern of cellular differentiation", J. Theor. Bio. 25:1.

Wolpert, L. (1971), "Positional information and pattern formation", in Current Topics in Developmental Biology 6, ed. A. A. Moscona and A. Monroy (New York: Academic Press).

Wood, E. H. (1976), "Cardiovascular and pulmonary dynamics by quantitative imaging", Circ. Res. 38:131.

Yates, F. (1950), "The place of statistics in the study of growth and form", Proc. Roy. Soc. Lond. B 137:479.

Young, R. W. (1956), "The measurement of cranial shape", Am. J. Phys. Anth. n.s. 14:59.

Zahn, C. T., and R. Z. Roskies (1972), "Fourier descriptors for plane closed curves", I.E.E.E. Trans. Comp. C-21:269.

Zeeman, E. C. (1974), "Primary and secondary waves in developmental biology", in Some Mathematical Questions in Biology. VI (Providence, R.I.: American Mathematical Society).

Zusne, L. (1970), Visual Perception of Form (New York: Academic Press).

Bio—
mathematics

Managing Editors: K. Krickeberg, S. A. Levin

Editorial Board: H. J. Bremermann, J. Cowan,
W. M. Hirsch, S. Karlin, J. Keller, R. C. Lewontin,
R. M. May, J. Neyman, S. I. Rubinow, M. Schreiber,
L. A. Segel

Volume 1:
Mathematical Topics in Population Genetics
Edited by K. Kojima
1970. 55 figures. IX, 400 pages
ISBN 3-540-05054-X

"...It is far and away the most solid product I have
ever seen labelled biomathematics."
American Scientist

Volume 2: E. Batschelet
Introduction to Mathematics for Life Scientists
2nd edition. 1975. 227 figures. XV, 643 pages
ISBN 3-540-07293-4

"A sincere attempt to relate basic mathematics to the
needs of the student of life sciences."
Mathematics Teacher

M. Iosifescu, P. Tăutu
**Stochastic Processes and Applications in Biology
and Medicine**

Volume 3
Part 1: **Theory**
1973. 331 pages.
ISBN 3-540-06270-X

Volume 4
Part 2: **Models**
1973. 337 pages
ISBN 3-540-06271-8

Distributions Rights for the Socialist Countries:
Romlibri, Bucharest

"... the two-volume set, with its very extensive bibliography, is a survey of recent work as well as a text-
book. It is highly recommended by the reviewer."
American Scientist

Volume 5: A. Jacquard
The Genetic Structure of Populations
Translated by B. Charlesworth, D. Charlesworth
1974. 92 figures. XVIII, 569 pages
ISBN 3-540-06329-3

"...should take its place as a major reference work.."
Science

Volume 6: D. Smith, N. Keyfitz
Mathematical Demography
Selected Papers
1977. 31 figures. XI, 515 pages
ISBN 3-540-07899-1

This collection of readings brings together the major
historical contributions that form the base of current
population mathematics tracing the development of
the field from the early explorations of Graunt and
Halley in the seventeenth century to Lotka and his
successors in the twentieth. The volume includes
55 articles and excerpts with introductory histories
and mathematical notes by the editors.

Volume 7: E. R. Lewis
Network Models in Population Biology
1977. 187 figures. XII, 402 pages
ISBN 3-540-08214-X

Directed toward biologists who are looking for an
introduction to biologically motivated systems
theory, this book provides a simple, heuristic
approach to quantitative and theoretical population
biology.

Springer-Verlag
Berlin
Heidelberg
New York

A
Springer
Journal

Journal of

Mathematical Biology

Ecology and Population Biology
Epidemiology
Immunology
Neurobiology
Physiology
Artificial Intelligence
Developmental Biology
Chemical Kinetics

Edited by H.J. Bremermann, Berkeley, CA; F.A. Dodge, Yorktown Heights, NY; K.P. Hadeler, Tübingen; S.A. Levin, Ithaca, NY; D. Varjú, Tübingen.

Advisory Board: M.A. Arbib, Amherst, MA; E. Batschelet, Zürich; W. Bühler, Mainz; B.D. Coleman, Pittsburgh, PA; K. Dietz, Tübingen; W. Fleming, Providence, RI; D. Glaser, Berkeley, CA; N.S. Goel, Binghamton, NY; J.N.R. Grainger, Dublin; F. Heinmets, Natick, MA; H. Holzer, Freiburg i. Br.; W. Jäger, Heidelberg; K. Jänich, Regensburg; S. Karlin, Rehovot/Stanford CA; S. Kauffman, Philadelphia, PA; D.G. Kendall, Cambridge; N. Keyfitz, Cambridge, MA; B. Khodorov, Moscow; E.R. Lewis, Berkeley, CA; D. Ludwig, Vancouver; H. Mel, Berkeley, CA; H. Mohr, Freiburg i. Br.; E.W. Montroll, Rochester, NY; A. Oaten, Santa Barbara, CA; G.M. Odell, Troy, NY; G. Oster, Berkeley, CA; A.S. Perelson, Los Alamos, NM; T. Poggio, Tübingen; K.H. Pribram, Stanford, CA; S.I. Rubinow, New York, NY; W.v. Seelen, Mainz; L.A. Segel, Rehovot; W. Seyffert, Tübingen; H. Spekreijse, Amsterdam; R.B. Stein, Edmonton; R. Thom, Bures-sur-Yvette; Jun-ichi Toyoda, Tokyo; J.J. Tyson, Blacksbough, VA; J. Vandermeer, Ann Arbor, MI.

Springer-Verlag
Berlin
Heidelberg
New York

Journal of Mathematical Biology publishes papers in which mathematics leads to a better understanding of biological phenomena, mathematical papers inspired by biological research and papers which yield new experimental data bearing on mathematical models. The scope is broad, both mathematically and biologically and extends to relevant interfaces with medicine, chemistry, physics and sociology. The editors aim to reach an audience of both mathematicians and biologists.